T0140039

Studies in Computational Intelligence

Volume 917

Series Editor

Janusz Kacprzyk, Polish Academy of Sciences, Warsaw, Poland

The series "Studies in Computational Intelligence" (SCI) publishes new developments and advances in the various areas of computational intelligence—quickly and with a high quality. The intent is to cover the theory, applications, and design methods of computational intelligence, as embedded in the fields of engineering, computer science, physics and life sciences, as well as the methodologies behind them. The series contains monographs, lecture notes and edited volumes in computational intelligence spanning the areas of neural networks, connectionist systems, genetic algorithms, evolutionary computation, artificial intelligence, cellular automata, self-organizing systems, soft computing, fuzzy systems, and hybrid intelligent systems. Of particular value to both the contributors and the readership are the short publication timeframe and the world-wide distribution, which enable both wide and rapid dissemination of research output.

Indexed by SCOPUS, DBLP, WTI Frankfurt eG, zbMATH, SCImago.

More information about this series at http://www.springer.com/series/7092

Huimin Lu
Editor

Artificial Intelligence and Robotics

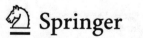 Springer

Editor
Huimin Lu
Department of Mechanical and Control
Engineering
Kyushu Institute of Technology
Kitakyushu, Japan

ISSN 1860-949X   ISSN 1860-9503   (electronic)
Studies in Computational Intelligence
ISBN 978-3-030-56180-2   ISBN 978-3-030-56178-9   (eBook)
https://doi.org/10.1007/978-3-030-56178-9

© Springer Nature Switzerland AG 2021
This work is subject to copyright. All rights are reserved by the Publisher, whether the whole or part of the material is concerned, specifically the rights of translation, reprinting, reuse of illustrations, recitation, broadcasting, reproduction on microfilms or in any other physical way, and transmission or information storage and retrieval, electronic adaptation, computer software, or by similar or dissimilar methodology now known or hereafter developed.
The use of general descriptive names, registered names, trademarks, service marks, etc. in this publication does not imply, even in the absence of a specific statement, that such names are exempt from the relevant protective laws and regulations and therefore free for general use.
The publisher, the authors and the editors are safe to assume that the advice and information in this book are believed to be true and accurate at the date of publication. Neither the publisher nor the authors or the editors give a warranty, expressed or implied, with respect to the material contained herein or for any errors or omissions that may have been made. The publisher remains neutral with regard to jurisdictional claims in published maps and institutional affiliations.

This Springer imprint is published by the registered company Springer Nature Switzerland AG
The registered company address is: Gewerbestrasse 11, 6330 Cham, Switzerland

Preface

In August 2019, the 4th International Symposium on Artificial Intelligence and Robotics took place in Daegu, Korea. This conference was organized by the International Society for Artificial Intelligence and Robotics (ISAIR, https://isair. site), Daegu Convention & Visitors Association, and Kyushu Institute of Technology, Japan. The annual organized series conferences focus on the exchanges of new ideas and new practices in industry applications. This book's objective is to provide a platform for researchers to share their thoughts and findings on various issues involved in artificial intelligence and robotics.

The integration of artificial intelligence and robotics technologies has become a topic of increasing interest for both researchers and developers from academic fields and industries worldwide. It is foreseeable that artificial intelligence will be the main approach of the next generation of robotics research. The explosive number of artificial intelligence algorithms and increasing computational power of computers has significantly extended the number of potential applications for computer vision. It has also brought new challenges to the artificial intelligence community. The aim of this book is to provide a platform to share up-to-date scientific achievements in this field. ISAIR 2019, had received over 280 papers from over 12 countries in the world. After the careful review process, 19 papers were selected based on their originality, significance, technical soundness, and clarity of exposition. The papers of this book were chosen based on review scored submitted by members of the program committee and underwent further rigorous rounds of review.

It is our sincere hope that this volume provides stimulation and inspiration, and that it will be used as a foundation for works to come.

Kitakyushu, Japan
May 2020

Editor
Huimin Lu

Acknowledgements

This book was supported by Leading Initiative for Excellent Young Research Program of Ministry of Education, Culture, Sports, Science and Technology of Japan (16809746), Strengthening Research Support Project of Kyushu Institute of Technology, Kitakyushu Convention & Visitors Association.

We would like to thank all authors for their contributions. The editors also wish to thank the referees who carefully reviewed the papers and gave useful suggestions and feedback to the authors. Finally, we would like to thank Prof. Hyoungseop Kim, Prof. Manu Malek, Prof. Pin-Han Ho, Prof. Geoffrey C. Fox, Prof. Kaori Yoshida, and all editors of Studies in Computational Intelligence for the cooperation in preparing the book.

About This Book

This edited book presents scientific results of the research fields of artificial intelligence and robotics. The main focus of this book is on the new research ideas and results for the mathematical problems in robotic vision systems.

In August 2019, the 4th International Symposium on Artificial Intelligence and Robotics (ISAIR 2019) took place in Daegu, Korea. This conference is an annual event sponsored by International Society for Artificial Intelligence and Robotics (ISAIR), International Association for Pattern Recognition (IAPR), and technology supported by IEEE Computer Society Big Data STC, and SPIE. ISAIR is to promote the research of the future information technology and trying to bring together the researchers from academia and industry to share and discuss ideas, problems and solutions in various areas of information technology, through series of activities such as conferences, symposium, and special sessions.

ISAIR2019 had received over 280 papers from over 12 countries in the world. The present book comprises selected contributions from this conference. The chapters were chosen based on review scores submitted by editors, and underwent further rigorous rounds of review. This publication captures 19 of the most promising papers, and we impatiently await the important contributions that we know these authors will bring to the fields of artificial intelligence and robotics.

Contents

CBCNet: A Deep Learning Approach to Urban Images Classification in Urban Computing

Zhenbing Liu, Zeya Li, Lingqiao Li, and Huihua Yang

Abstract Urban images classification is an import part of urban computing. It is a challenging task for object detection and classification of urban images due to the high complexity of image contents. i.e., an image may contain buildings, pedestrians, vehicles, roads, etc. In this paper, a novel convolutional neural network, named Complex Background Classification Network (CBCNet), is proposed for classifying the urban images. This network is unlike existing AlexNet, ResNet, etc. It uses a multilayer perceptron convolutional layer to extract more representative features of complex urban images, instead of using a linear convolutional layer, and integrates back-propagation network to optimize object parameters. We also build a standard dataset of urban images containing eight categories, contrast experiments prove that the dataset is rational and feasibility. Experiment results obtained on two benchmark datasets demonstrate that classification accuracy and computation of CBCNet outperform those by the previous state-of-the-art items of AlexNet, VGGNet16 and ResNet50.

Keywords Classification · Urban computing · Dataset · Object detection · Deep learning

Z. Liu · Z. Li · L. Li (✉) · H. Yang
School of Computer Science and Information Security, Guilin University of Electronic Technology, Guilin 541004, China
e-mail: 54pe@163.com

Z. Li
e-mail: Zyli2019@163.com

L. Li · H. Yang
School of Automation, Beijing University of Posts and Telecommunications, Beijing 100876, China

© Springer Nature Switzerland AG 2021
H. Lu (ed.), *Artificial Intelligence and Robotics*,
Studies in Computational Intelligence 917,
https://doi.org/10.1007/978-3-030-56178-9_1

1 Introduction

The arrival of information age brings urbanization development into a digital, smart and mobile stage. Report of the major issues that cities face will depend more on information technologies like smart phones and cameras [1, 2]. In China, now most of urban management systems are running efficiently, citizens can use mobile APP, WeChat client, computer and other terminals to upload images and fill out relevant information (i.e., location, main category, real situation, etc.) when they find and report the urban issues [3, 4]. Then, urban management staffs delegate the issues to the related administration according to experience and providing information. However, since the continued expansion of cities, the urban issues have been increasing so rapidly that manual distributions of the administration are unable to meet the needs of daily work. Hence, there urgently needs a quick classification tool of urban issues to improve the efficiency of urban management and make it intelligentized. Finally, in this paper, the problem of image classification is transformed into an image recognition task, and we propose a deep learning approach to the classification of high definition urban images for processing efficiency in urban management system.

Extracting image features is making good progress from artificial feature to deep learning classification methods which the higher-level features are represented by combination of lower-level features of the data. Some traditional classification methods have been proposed from different perspectives, such as using Gabor wavelet image [5], Gaussian Markov random filed (GMRF) [6] and Scale-invariant feature transform (SIFT) [7] to extract textural features, remote sensing image features and digital image features, then using support vector machine (SVM) [8], Lagrange support vector machine (LSVM) [9] classification method and other traditional machine learning methods to classify images. In 1962, Hubel and Wiesel [10] presented the concept of receptive field. In 1980, Fukelima [11] proposed a neural cognitive machine, which is the first network of CNN. In 1989, Lecun et al. [12] proposed the back-propagation algorithm, which promoted the development of convolutional neural networks greatly. In 1998, Lecun et al. proposed the structure of the LeNet-5 network [13]. This basic design has good performance on MNIST, CIFAR and other data sets, especially on ImageNet Category Challenge [14, 15]. From 2012 to 2015, with development of computer hardware and intensive research on convolutional neural networks, deeper and more complete network models have been breaking records on ImageNet, such as AlexNet by Krizhevsky et al. [15], VGGNet of Oxford University [16], Google's GoogleNet [17] and Microsoft's ResNet [18].

Urban images are collected by citizens and outdoor workers who take urban issues photos by their own ordinary mobile phone. The collected image exists drawbacks, its complex background information, low-resolution and uneven brightness make it difficult to obtain exact results. Since urban images in urban management system are manually collected, they must contain objects which represent key information and vital features. If those key objects can be detected from the complex background of the image, it is equivalent to extract the critical features that can represent image

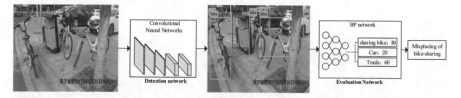

Fig. 1 Framework of the CBCNet classification

information, which can significantly improve the classification performance of the model. In this paper, we build a standard urban images dataset containing eight categories, including Graffiti and Posters, Gutter cover plate damage, etc. Through screening, sorting and categorization, we build this hierarchical dataset. To overcome these above-mentioned problems, we proposed a new convolutional neural network, named complex background classification network (CBCNet), classifying the urban images. The overview of the CBCNet classification framework for urban images is shown in Fig. 1, Processing images with CBCNet is simple and straightforward.

Our network draws on the idea that image classification to predict all information may depend on a key object. Instead of using region proposal networks method, such as R-CNN [19], Faster R-CNN [20], which is slow and difficult to optimize, the entire image is used as the input with end-to-end approach, information of available key objects can be directly got by optimizing the convolution model [21]. In this paper, extensive experiments are conducted on the urban images dataset we build and the PASCAL VOC 2007 dataset to verify effectiveness of this method.

The remainder of this paper is organized as follows. In Sect. 2, CBCNet model is introduced in detail. The experimental results and analysis are presented in Sect. 3, followed by the conclusion in Sect. 4.

2 Proposed Method

2.1 Detection Network

The detection part of CBCNet model is designed according to the complex background and multi scale images. We use 1×1 reduction layers followed by every 3×3 convolutional layers to abstract representative data within the receptive filed [22]. The full network is shown in Fig. 2.

Convolutional layers are employed to extract image feature. Each layer uses an established kernel to convolute the image, adds weights and biases to the convolution value, and then acts on an activation function. Because the convolutional kernel used in this paper is relatively small, traversal of matrices is used to implement the convolution operation. The concrete flow of detection network is: (1) Processing the image data into a size of 448 * 448 as an input to the model; (2) Setting the

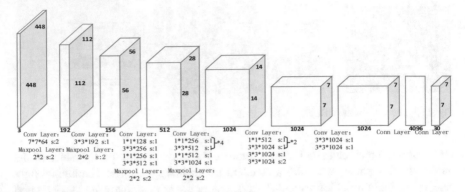

Fig. 2 Detection network structure of CBCNet

convolution kernel size to 7 * 7, the convolution step size to 2, using 64 filters for convolution of the input image, using Rectified Linear Unit Function (ReLU) as the activation function in the convolution calculation, and Batch Normalization (BN) is added before every conv-layer to accelerate convergence and regularization; (3) Using max-pooling, selecting the maximum value from every four pixels in the feature map; (4) Adding multilayer perceptron to break the dimension reduction module; (5) Finally, a 7 * 7 * 18 vector result is output, where 7 * 7 is the 49 grids into which the image is divided, and 18 dimensions is the prediction result in each grid, where 10 dimensions are the coordinates of the key object position, 8 dimension is the probability of the urban cases categories to which the key object belongs.

In order to allow the network to extract more abstract and effective non-linear features, part of convolutional layers is substituted with multilayer perceptron (MLP) instead linear convolutional layer. The convolutional layer usually applies linear convolution using a linear filter followed by a nonlinear activation function to generate a feature map, but if all possible features are extracted with a super-complete filter, the network will too big. Therefore, the network adds a micro neural network to each local receptive field in the convolution operation [22], as shown in Figs. 3 and 4.

In this paper, cascading cross channel parametric pooling layer, when the MLP convolutional layers is applied to convolutional layer, this method can be treated as an additional 1 × 1 convolutional layer between two conv-layers, followed by a

Fig. 3 Linear conv-layer

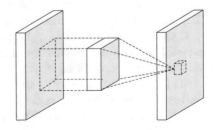

Fig. 4 Optimized conv-layer

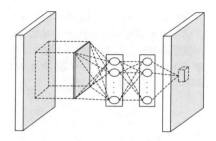

leaky rectified linear activation, as shown in (1). However, it has dual purpose: the most important point is that they are mainly used for dimension reduction modules to break calculation bottleneck, otherwise size of our network will be limited, it exerts a vital influence on today's work efficiency of urban management system. This allows us not only to deepen the network and widen our network without causing serious performance degradation, but also to accelerate network optimization. We can find sometimes the overparameterization is possibly useful.

$$\varphi(x) = \begin{cases} x & \text{if } x > 0 \\ 0.1x & \text{otherwise} \end{cases} \tag{1}$$

2.2 Evaluation Network

The evaluation part of CBCNet model is designed by back-propagation neural network, as is shown in Fig. 5, which belongs to multi-layer feed-forward artificial neural network, which consists of several neurons, and each neuron corresponds to an activation function and a threshold. If the output layer cannot get the desired output, the error signal is propagated back layer by layer, and the error is reduced by modifying weights and bias between the layers.

Fig. 5 Evaluation network structure of CBCNet

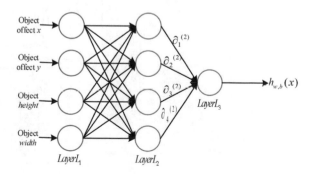

In this paper, we use the width w, the height h and the center relative offset x, y of the single key object contained in each urban image as the input vector, by using forward propagation of the alternating input and back propagation of the error, making the change of weights become little by gradient descent method, finally it attains the minimal error.

3 Training

We use urban images dataset with annotated object to train detection network to extract information of urban images' object contained, as shown in Fig. 6. We use data to retrain CBCNet model which based on the YOLO model. Relationship between image issues category and the annotated object is used to train the evaluation network to get entire correct classification, and hence the entire network training is completed. When training the detection model, it is easy to cause over-fitting due to small sample in training set, we use data augmentation to expand dataset through fine-tuning of rotation, exposure, saturation and tone of original image.

The detection model of CBCNet model uses sum-squared error as loss function to optimize parameters, that is sum-squared of network output vector and the real image corresponding vector, as shown in formula 2, where and represent the predicted data and the annotated data. In this paper, contribution of object position's correlated errors and category errors to the loss value in this network is different. Therefore, when calculating the loss, we set $\lambda_{coord} = 6.2$ to increase position's contribution and $\lambda_{noop} = 0.4$ to reduce influence of the grid which doesn't contain object.

As shown in formula 3, it optimizes multi-part loss function when training the model. Where x and y represent offset of the center of bounding box from grid boundary, w and h represent ratio of bounding box's width and height to the original

Fig. 6 Illustration of training process for CBCNet

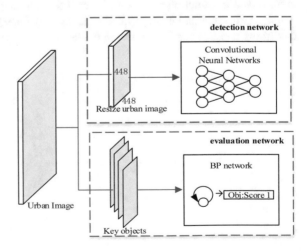

image's, $p(C)$ represents predicted category accuracy of C, $C = 8$ represents eight categories contained in dataset. Where Π_i^{obj} denotes if object appears in cell i, Π_i^{obj} denotes that the jth bounding box predictor in cell i is "responsible" for the prediction. We predict multiple bounding boxes per grid cell. In training section, we only want one bounding box predictor to be responsible for each object [23]. A three layers back-propagation neural network is used when training classification network. w, h and x, y in each image are used as input vector. We treat objects that affect classification as positive sample and irrelevant objects as negative sample. Through the input and back propagation of the error, the minimum value of error function is achieved by iteratively searching for weight vector using gradient descent method.

$$SSE = \sum_{i=0}^{n} \lambda_i (y_i - \hat{y}_i)^2 \qquad (2)$$

$$Loss = \lambda_{coord} \sum_{i=0}^{S^2} \sum_{j=0}^{B} \Pi_{i,j}^{obj} \left[(x_i - \hat{x}_i)^2 + (y_i - \hat{y}_i)^2 \right]$$

$$+ \lambda_{coord} \sum_{i=0}^{S^2} \sum_{j=0}^{B} \Pi_{i,j}^{obj} \left[(\sqrt{w_i} - \sqrt{\hat{w}_i})^2 + (\sqrt{h_i} - \sqrt{\hat{h}_i})^2 \right]$$

$$+ \sum_{i=0}^{S^2} \sum_{j=0}^{B} \Pi_{i,j}^{obj} (C_i - \hat{C}_i)^2 \qquad (3)$$

$$+ \lambda_{noob} \sum_{i=0}^{S^2} \sum_{j=0}^{B} \Pi_{i,j}^{no-obj} (C_i - \hat{C}_i)^2$$

$$+ \sum_{i=0}^{S^2} \Pi_i^{obj} \sum_{c \in classes} (p_i(c) - \hat{p}_i(c))^2$$

4   Experiments and Results

4.1   Dataset Description and Experimental Design

Urban image dataset is based on the intelligent management of urban management system. Based on urban issues statistics information in current actual operation, dataset is concentrated, categories are summarized and hierarchical category structure of the dataset is constructed, the specific sample images are shown in Fig. 7. Data is selected from four main categories (Urban Environment, Urban Appearance, Urban Facilities, Urban Traffic) and eight sub-categories, respectively: Graffiti and Posters, Gutter cover plate damage, Manhole cover damage, Traffic

Fig. 7 Sample images of the urban images dataset

guardrail damage, Misplacing of bike-sharing, Misplacing of non-motor vehicle, Illegal parking of motor vehicle, and Exposed garbage. There are 4119 pictures of all sizes in dataset with high definition image format, the size of entire dataset is about 500 M. The ratio of training set and test set is 3:2. Each category starts from '000001' with suffix and increases in turn. The specific dataset information is shown in Table 1.

$$Accuracy = \frac{TP}{S} \times 100\% \qquad (4)$$

Table 1 Category of urban image dataset

Class	Training set	Testing set	Total
Graffiti and Poster	375	174	549
Gutter cover plate damage	292	260	552
Manhole cover damage	391	186	567
Traffic guardrail damage	318	152	470
Misplacing of bike-sharing	226	181	407
Misplacing of non-motor vehicle	321	204	525
Illegal parking of motor vehicle	300	245	545
Exposed garbage	296	208	504
Total	2519	1610	4119

Table 2 Comparison results of object detection on urban images dataset

Model	Accuracy (%)
YOLO	96.32
Faster-RCNN	**96.93**

YOLO and Faster-RCNN are selected to classify urban images. The experiments are based on the deep learning framework Darknet in the hardware environment configured as Ubuntu14.04, the graphics card is M40, and CPU is Intel XeonE5. We evaluate effect of the model by calculating classification accuracy of each category, as shown in formula 4, where *TP* is the true positive, *S* is total number of all category. Both YOLO algorithm and Faster-RCNN algorithm have achieved good results. Among them, Faster-RCNN reached 96.93%, which was about 0.6% higher than YOLO (Table 2).

4.2 Classification Experiments Based on CBCNet

Experiment is executed using same above-mentioned environment with urban images dataset. In our experiments, we use precision for evaluating our model's performance, it is the number of correctly labels divided by the number of generated labels, here is formula 5 of calculation precision. Where *TP* means true positives, *FP* means false positives. We make use of *Mean_pre* to evaluate model's classification performance, which can be achieved as follow formula 6. where *n* means the total of all categories in our dataset.

$$Precision = \frac{TP}{TP + FP} \times 100\% \tag{5}$$

$$Mean\_pre = \frac{1}{n} \sum_{i=1}^{n} Precision_i \tag{6}$$

By calculating and comparing classification phenomenon in urban image dataset with AlexNet [15], VGGNet16 [16], ResNet [18] and CBCNet, the experiment results are shown in Tables 3 and 4, it shows that our network has average precision of 97.23% in test set. The category with the highest classification precision 'Graffiti and Poster' is 99.87% and the lowest 'Manhole cover damage' is 87.62%. It can be seen that CBCNet is superior to AlexNet, VGGNet16 and ResNet in classifying urban images, which mainly manifest in detection speed and classification accuracy. In the almost same training time, AlexNet has similar classification result with CBCNet method, but there are considerable differences in precision of 'Gutter cover plate damage', 'Illegal parking of motor vehicle' and 'Misplacing of non-motor vehicle', images of 'Misplacing of non-motor vehicles' has more complex background than 'Graffiti and Posters', since street images are more complex than image of a wall. On cases with

Table 3 Experiment results on each category of urban images dataset

Category	Graffiti and Posters (%)	Gutter cover plate damage (%)	Manhole cover damage (%)	Traffic guardrail damage (%)	Misplacing of sharing bicycles (%)	Misplacing of non-motor vehicles (%)	Illegal parking of motor vehicles (%)	Exposed garbage (%)
AlexNet	95.05	90.16	69.91	94.73	93.92	79.65	98.48	97.42
Vgg16Net	91.70	97.70	**94.50**	92.77	**98.84**	96.13	99.15	**99.01**
Res50Net	95.02	91.47	79.80	90.90	96.59	87.03	98.23	96.96
CBCNet	**99.87**	**99.31**	87.62	**95.52**	**98.84**	**98.85**	**99.16**	98.57

Table 4 Experiment results of different models on urban images dataset

Model	Training time (h)	*Pre* (%)	*Recall* (%)	*F1_Score*
AlexNet	6	89.91	90.83	0.899
Vgg16Net	7	96.22	97.30	0.970
Res50Net	20	92.00	92.61	0.922
CBCNet	6	**97.23**	**98.42**	**0.973**

complex backgrounds like 'Illegal parking of motor vehicles' and 'Misplacing of non-motor vehicles', CBCNet still out performs better than VGGNet16. Comparing ResNet50 and CBCNet classification results for each case category, we find that CBCNet performs fully exceeds ResNet50.

4.3   Classification Experiment Based on VOC Dataset

The PASCAL VOC is a benchmark test for visual recognition and objects detection, it provides standard image annotation dataset and evaluation systems, which contains a total of 9963 marked pictures, divided into 20 categories, consists train/validation/test three parts, and has 24,640 annotated objects. In our experiment, we use the images only contain a single object as training set, others as testing set, to compare with AlexNet, VGGNet16 and ResNet50 [24]. As is shown in Table 5, the highest classification accuracy is ResNet50 network with 73.8%, and our method is the second but also reach to 72.4%. Using CBCNet, class 'cat' has the highest classification accuracy 89.6%, class 'potted plant' has the lowest classification accuracy 48.1%.

Table 5 Experimental results on VOC standard dataset

Model name	AlexNet (%)	VGGNet (%)	ResNet (%)	**CBCNet** (%)
Mean_pre	65.4	71.5	**73.8**	72.4

5 Conclusion

To handle the increasing number of cases for urban management efficiently in urban computing, in this paper we have proposed a deep learning approach for complex background urban images classification, which is able to not only classify cases based on urban images, but also improve efficiency of urban management. In framework of the CBCNet classification, this method classifies issues by detecting information of key objects in urban image, and makes up low accuracy of current images classification algorithm in complex background images. We use urban images and test our method's classification performance from various aspects. The experimental data shows that this method can classify urban images automatically, with 97.23% accuracy. In the same hardware environment, this method can still consider good classification accuracy while guaranteeing training speed compared to AlexNet, VGGNet, ResNet. In the future work, we will further improve and open our dataset, combined with algorithms such as image segmentation and image denoising to improve classification accuracy and efficiency of the urban management system.

Acknowledgements This study is supported by the National Natural Science Foundation of China (Grant No. 61562013, 61906050), the Natural Science Foundation of Guangxi Province (CN) (2017GXNSFDA198025), the Key Research and Development Program of Guangxi Province under Grant (AB16380293, AD19245202), the Study Abroad Program for Graduate Student of Guilin University of Electronic Technology (GDYX2018006).

References

1. Yang C (2014) Research on construction of digital intelligent city management system. Int J Hybrid Inf Technol
2. Zheng Y, Mascolo C, Silva CT (2017) Guest editorial: urban computing. IEEE Trans Big Data 3(2):124–125
3. Zhang BX, Wang XC (2014) The reality of plight faced by city management. Urban Probl 05:79–84
4. Dai P, Jing C, Du M et al (2006) A method based on spatial analyst to detect hot spot of urban component management events. In: IEEE international conference on spatial data mining and geographical knowledge services, vol 8, no 10, pp 55–59, Jan 2006
5. Moghaddam HA, Saadatmand-Tarzjan M (2006) Gabor wavelet correlogram algorithm for image indexing and retrieval. In: International conference on pattern recognition, vol 2, pp 925–928
6. Seetharaman K (2015) Image retrieval based on micro-level spatial structure features and content analysis using full range Gaussian Markov random field model. Eng Appl Artif Intell 13(40):103–116
7. Lindeberg T (2012) Scale invariant feature transform. Scholarpedia 5(1):2012–2021
8. Adankon MM, Cheriet M (2009) Support vector machine. In: International conference on intelligent networks and intelligent systems, pp 418–421
9. Hwang JP, Choi B, Hong IW et al (2013) Multiclass Lagrangian support vector machine. Neural Comput Appl 3(4):703–710
10. Hubel DH, Wiesel TN (1962) Receptive fields, binocular interaction and functional architecture in the cat's visual cortex. J Physiol 160(1):151–160

11. Fukushima K (1980) Neocognitron: a self-organizing neural network model for a mechanism of pattern recognition unaffected by shift in position. Biol Cybern 36(4):193–202
12. Lecun Y, Boser B, Denker J et al (1989) Backpropagation applied to handwritten zip code recognition. Neural Comput 1(4):541–551
13. Lécun Y, Bottou L, Bengio Y et al (1998) Gradient-based learning applied to document recognition. Proc IEEE 86(11):2278–2324
14. Russakovsky O et al (2015) ImageNet large scale visual recognition challenge. Int J Comput Vis 115(3):211–252
15. Krizhevsky A, Sutskever I, Hinton GE (2012) ImageNet classification with deep convolutional neural networks. In: International conference on neural information processing systems, pp 1097–1105
16. Simonyan K, Zisserman A (2014) Very deep convolutional networks for large-scale image recognition. Comput Sci 1409–1556
17. Szegedy C, Liu W, Jia Y et al (2015) Going deeper with convolutions. Comput Vis Pattern Recogn 1(9):1409–4842
18. He K, Zhang X, Ren S et al (2016) Deep residual learning for image recognition. Comput Vis Pattern Recogn 770–778
19. Girshick R, Donahue J, Darrell T et al (2014) Rich feature hierarchies for accurate object detection and semantic segmentation. Comput Vis Pattern Recogn 11(23):580–587
20. Ren S, He K, Girshick R et al (2017) Faster R-CNN: towards real-time object detection with region proposal networks. IEEE Trans Pattern Anal Mach Intell 39(6):1137–1149
21. Cengil E, Cinar A et al (2017) A GPU-based convolutional neural network approach for image classification. Intell Data Process Symp 16(17):1–6
22. Lin M, Chen Q, Yan S (2013) Network in network. Comput Sci 10(4)
23. Redmon J, Divvala S, Girshick R et al (2015) You only look once: unified, real-time object detection. Comput Vis Pattern Recogn 27(30):779–788
24. Zhong G et al (2017) Reducing and stretching deep convolutional activation features for accurate image classification. Cogn Comput 10(10):1–8

Building Label-Balanced Emotion Corpus Based on Active Learning for Text Emotion Classification

Xuefeng Shi, Xin Kang, Ping Liao, and Fuji Ren

Abstract In Supervised-learning of emotions from human language, to keep the emotion labels balanced in the training set is a challenging task since emotion labels are highly biased in raw data of human language. In this paper, we propose a novel method based on active learning to partially inhibit the polarization of text samples with more frequently observed emotion labels for constructing the training set, and to encourage the selection of samples with less frequently observed emotion labels. For each batch of unlabeled samples, the selected samples by our approach are given the ground truth emotion labels from human experts before they are merged to the training data. Our experiment of multi-label emotion classification on Chinese Weibo messages suggests that the proposed method is effective in constructing the label-balanced training set for text emotion classification, and the supervised text emotion classification results have been steadily improved with such training set.

Keywords Multi-label emotion classification · Active learning · Label balancing

X. Shi · P. Liao
School of Mechanical Engineering, Nantong University, No. 9 Seyuan Road, Nantong, Jiangsu, China
e-mail: sxfmch@163.com

P. Liao
e-mail: liao.p@ntu.edu.cn

X. Shi · X. Kang · F. Ren (✉)
Faculty of Engineering, Tokushima University, 2-1 Minamijyousanjima-cho, Tokushima 770-8506, Japan
e-mail: ren@is.tokushima-u.ac.jp

X. Kang
e-mail: kang-xin@is.tokushima-u.ac.jp

© Springer Nature Switzerland AG 2021
H. Lu (ed.), *Artificial Intelligence and Robotics*,
Studies in Computational Intelligence 917,
https://doi.org/10.1007/978-3-030-56178-9_2

1 Introduction

Emotions have been found useful for clarifying the mental thoughts underlying huge number of social network messages for solving an increasing number of the real-world problems, such as public opinion analysis [1–3], disease diagnosis [4–6], stock trend prediction [7–9], and product review evaluation [10–12]. Correctly understanding the emotional information through social network messages helps the analysis of future trends in these fields and provides precious guidance for the decision making at once.

Emotion classification research focuses on the recognition of delicate human emotions [13, 14], which is different from the sentiment classification for the positive and negative emotion polarities. There has been tiny variance of the human emotion definitions in different research fields. For example, Ekman [15] proposed six basic human emotion categories of Anger, Disgust, Fear, Happiness, Sadness, Surprise for facial emotion recognition, while Ren et al. [24] proposed eight basic emotion categories of Anger, Joy, Sorrow, Anxiety, Hate, Expect, Surprise, and Love for language emotion classification. In this paper, we would employ the eight basic emotion categories from [24] for studying the social network language emotion classification methods.

One of the most significant problems in learning emotion classifiers for natural language is that emotion labels are distributed in a highly biased manner in the raw data. This makes a training set of well-balanced emotion labels very difficult to build based on the real-world raw data and restricts the learned emotion classifiers for recognizing the less frequent emotion labels, such as anxiety and surprise. In this paper, we propose a novel method to actively select samples which are potentially indicative of the less frequent emotion labels from a huge number of social network raw messages, for building a balanced training set for emotion classification.

Specifically, given an existing training set of unbalanced emotion labels and a set of raw text samples, our algorithm at first generates the probabilistic predictions for all the raw samples and temporarily merge their probabilistic emotion predictions into the existing training set to construct a bunch of candidate training sets. Then by explicitly evaluating the Kullback–Leibler divergence of the emotion label distributions in different candidate training sets to the ideal uniform emotion distribution, the algorithm incrementally finds the most promising candidate training sets as well as the samples with the most promising label-balancing property accordingly. Finally, the samples are given ground truth emotion labels by human experts and could be merged to the training data permanently.

The rest of the paper is arranged as follows. We review the related work in Sect. 2. In Sect. 3, we describe the algorithm for constructing a label-balanced training set for text emotion classification. Section 4 describes the experiment set up and analyses the text emotion classification results. Our conclusions and future work would be given in Sect. 5.

2 Related Work

Due to the requirement of different applications, text emotion recognition can be divided into three granularities: the tagging of word emotions, the classification of sentence emotions, and the analysis of document emotions [16]. The word emotion tagging mainly focus on finding the emotional words from documents and predicting the emotions in these words [17–19]. The analysis of document emotions focuses on the identification of emotional factors in articles and a further prediction of the emotions in the document through various machine learning methods [20]. The classification of sentence emotion depends on the analysis of word emotions within the short-to-middle length texts, and provides abundant information for the document emotion analysis. Previous works [21–23] suggested that a sentence could have more than one emotion states at the same time, which indicates a multi-label classification problem for sentence emotion classification. In this work, we train the multi-label emotion classifiers for the short-to-middle length social network messages.

Ren et al. [24] employed the Hierarchical Bayesian network to generate the latent topics and emotion labels based on the assumption of probabilistic dependencies over words, topics, and emotion labels, for predicting the complex human emotions in documents. By analyzing the distribution of emotion labels and topics, they found the variation of emotional thoughts towards different semantic topics. Liu et al. [25] employed the large-scale real-world knowledge for sensing the textual affect, which contained inherent affective information. Their method was robust enough on predicting the affective qualities of the underlying semantic content of text based on the understanding of real-world knowledge. Because the existing active learning methods could not get the examples which were the most informative and representative, Reyes et al. [26]. proposed a new active learning sample selection strategy based on the predictions of a base classifier and the inconsistency of a predicted label set. The proposed method got better results than several state-of-the-art strategies on a large number of datasets. Li et al. [27] proposed a method based on the measure of the candidate example's uncertainty, which took the maximum margin between the predictions from a set of binary SVM classifiers, and a Label Cardinality Inconsistency (LCI) strategy which controlled the rate of prediction mistake. Kang et al. [28] proposed four criteria which evaluated the complementariness, informativeness, representativeness, and diverseness of the candidates for sample selection in active learning. The examples selected by this method could progressively improve the supervised classification result when they were complemented into the training set. All these works have focused on the improvement of prediction accuracy for the selected examples, but few works of active learning were done for inhibiting the label polarization in the training set after the selected samples were complemented.

3   Balanced Emotional Corpus Construction

In this part, we would illustrate our platform in detail for constructing a well-balanced training set of emotional samples and for learning a multi-label emotion classifier. These correspond to an active learning model as an extension of Kang et al.'s work [28] and a basic multi-label emotion classification model, respectively. In the active learning model, we extend Kang et al.'s work [28] by modifying the order of sample selection procedure and by explicitly evaluating the label-balancing property for each raw sample to improve the label balance in training set construction. The detailed descriptions of two models are as follows.

3.1   Multi-label Emotion Classification

In the multi-label emotion classification model, a series of logistic regression classifiers φ_k are constructed, each of which can learn to predict a probability $y_k \in [0, 1]$ for the emotion label k, given a Weibo message x as

$$y_k = \varphi_k(x). \tag{1}$$

We extract the bag-of-word features from message texts with the Chinese morphological analysis engine Thulac, with low frequency words and stop words filtered out. The result feature vector x indicates the observation of each word in a Weibo message. Hyper-parameters of the logistic regression classifiers including the l1 and l2 penalty, regularization strength, and class weights for each classifier φ_k, are selected through the 5-fold cross validation on the training set.

3.2   Active Learning Progress

As in the active learning model [28], text samples with the most significant informativeness and representativeness scores from the large set of raw data are greedily selected to be appended to the existing training set, which is then employed for updating the learning of the emotion classifiers φ_k. While different from the active learning model [28], we reorder the sample selection procedure by putting the complementariness criterion for inhibiting biased emotion labels after the other three sample selection criteria, i.e. the informativeness criterion, the representativeness criterion, the diverseness criterion. This allows our approach to directly adjust the label-balancing property in the final output and put more weights to the complementariness criterion accordingly. Furthermore, we redesigned the complementariness criterion to explicitly evaluate the Kullback-Leibler divergence between the emotion distribution in a temporary training set to the ideal uniform emotion distribution, which evaluates the label-balancing property for the corresponding raw sample in a more explicit manner. The derivations for each criterion are as follows.

The informativeness criterion evaluates the maximum of the probabilistic emotion prediction entropy for all emotion categories by

$$i(x) = \max_{k \in \{1,...k\}} (-y_k \log y_k - (1 - y_k) \log(1 - y_k)), \tag{2}$$

in which y_k indicates the probabilistic emotion prediction for category k. By maximizing this criterion as in Algorithm I, our approach could find the candidate samples with large information, i.e. high uncertainty for prediction, in at least one dimension of its emotion predictions.

The representativeness criterion evaluates the average similarity for each text sample to all other text samples in the raw data by

$$r(x) = \frac{1}{|U|} \sum_{x\prime \in U} -\sqrt{x \cdot x - 2x \cdot x\prime + x\prime \cdot x\prime}, \tag{3}$$

in which we use U to indicate the set of all raw samples and employ the opposite of the Euclidean distance between two samples x and x' to indicate their linguistic similarity. By maximizing this criterion as in Algorithm I, our approach could find the candidate samples which are linguistically representative of many other samples in the raw set.

The diverseness criterion evaluates the minimum of the Euclidean distances between a raw sample to all selected training samples by

$$d(x) = \min_{x' \in X} \sqrt{x \cdot x - 2x \cdot x' + x' \cdot x'}, \tag{4}$$

in which we use X to indicate the set of message samples in the training data. By maximizing this criterion as shown in Algorithm I, our approach could find the candidate samples which are linguistically distinct to the already selected training samples.

We propose a novel complementariness criterion for inhibiting the selection of biased emotion labels into the training set, by constructing a series of temporal training sets $X \cup \{x\}$ each of which corresponds to taking a raw sample $x \in U$ into the existing training set X, and by explicitly evaluating the Kullback–Leibler Divergence between the emotion distribution $p\prime$ in a temporary training set and the ideal uniform emotion label distribution $u \sim \text{unif}\{1, K\}$ to find the one of the smallest Kullback–Leibler Divergence $c(x)$, i.e. a new training set with the most balanced emotion labels, by

$$c(x) = - \sum_{k=1}^{K} p\prime_k(x) \log\left(\frac{u_k}{p\prime_k(x)}\right), \tag{5}$$

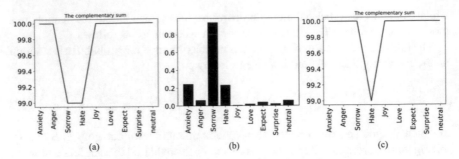

Fig. 1 In (**a**), the emotion labels in the training set are unbalanced for the short of Sorrow and Hate labels. In (**b**), the sample selection procedure for the complementary criterion finds a candidate sample with high probability of emotion label Sorrow. In (**c**), the unlabeled sample and its emotion label Sorrow is added to the training set, rendering a more balanced emotion label distribution

$$p'_k(x) = \frac{\sum_{x\prime \in X \cup \{x\}} e_k(x\prime)}{|X \cup \{x\}|}, \tag{6}$$

$$u_k = \frac{1}{K}, \tag{7}$$

in which $e_k(x)$ is the probability of observing emotion label k in sample x. For samples in the existing training set $x \in X$, the probability of observing emotion label k is either 1.0 or 0.0 given the ground truth emotion label. For samples in the raw message data $x \in U$, the probability of observing emotion label k is given by prediction result $e_k(x\prime) = \varphi_k(x\prime)$ from the logistic regression emotion classifier.

Figure 1 shows a concrete example of the sample selection procedure based on the complementariness criterion. With an emotion label distribution in the current training set X as shown in Fig. 1a, the proposed criterion tends to select a new sample x with probabilistic emotion labels as Fig. 1b from the raw message data U, which renders the temporary training set $X \cup \{x\}$ with a more balanced emotion label distribution, i.e. a smaller Kullback–Labler Divergence to the ideal uniform emotion label distribution, as shown in Fig. 1c.

Algorithm I. Building Label-Balanced Emotion Corpus through Active Learning
1.   Input: training set X, unlabeled set U, selection parameters λ
2. $I = \{i(x)|\forall x \in U\}$
3. $U^I = \text{argpartition}(I, \lambda^I|U|)^1$
4. $R = \{r(x)|\forall x \in U^I\}$
5. $U^R = \text{argpartition}(R, \lambda^R\lambda^I|U|)$
6. $D = \{d(x)|\forall x \in U^R\}$
7. $U^D = \text{argpartition}(D, \lambda^D\lambda^R\lambda^I|U|)$
8.   For $i = 0 \to \lambda^C$ do

9. $x = \operatorname{argmin}(\{c(x)|\forall x \in U^D\})$
10. Obtain emotion label e for x
11. Append (x, e) to X and delete x from U
12. End for

In Algorithm I, parameter values in $\lambda^I, \lambda^R, \lambda^D$ are the percentage ratios for sample selection based on the informativeness, representativeness, diverseness criteria, respectively. The fixed selection ratios in these parameters help maintaining the capabilities in these criteria for the sample selection given raw data sets of variable sizes. The parameter value in λ^C corresponds to the number of samples to be finally selected by our active learning algorithm. We employ a fixed number for the output size to make the learning procedure for text emotion classification comparable for different raw data sets in the experiment.

4 Experiment and Result

We carry out an emotion classification experiment for Weibo messages, and evaluate the emotion classification results with respect to different active learning algorithms. The initial training, validation, and test sets consist of 864, 1005, and 1592 Weibo messages, respectively. Each Weibo message has been annotated by human experts with one or more emotion labels from Joy, Love, Expectation, Surprise, Hate, Sorrow, Anger, Anxiety, and Neutral. The number of labels for each emotion category has been kept the same around, which corresponds to 100, 100, and 184 for the training, validation, and test sets, respectively. This is to allow our approach to learn an unbiased emotion classifier at the very beginning of active learning, and to evaluate the emotion classification results based on an even test embedding for each emotion category. The raw data has been randomly retrieved from Weibo stream and arranged into separate sets by the retrieval day and hour, with each set consists of several tens of thousands of Weibo messages.

We employ the validation set and 3 raw data sets to determine the selection parameters values in $\lambda^I, \lambda^R, \lambda^D$ and 6 raw data sets to determine the parameter value in λ^C for Algorithm I. Table 1 shows the candidate values for each selection parameter. Specifically, the candidate values for the first three parameters are percentage values for specifying the selection ratios, and the candidate values for the last parameter specify the final output size. By incrementally updating the training set for each group of parameter values, we collect and compare the micro and macro average F1 emotion classification scores on the validation set, and find the optimal parameter values of 0.2 for λ^I, 0.5 for λ^R, 0.5 for λ^D, and 40 for λ^C.

---

[1] The argpartition($F(X), n$) function selects n elements x from X which have larger scores $F(\cdot)$ than the other elements in X.

Table 1 Candidate values for the selection parameters in the active learning algorithm

Parameters	Candidate values
λ^I	0.50, 0.33, 0.25, 0.20
λ^R	0.50, 0.33, 0.25
λ^D	0.50, 0.33, 0.25
λ^C	20, 30, 40, 50

The experiment is conducted as follows. For each raw data set U, we firstly feed it together with the current training set X to Algorithm I, to get an updated training set. Then, we train a series of emotion classifiers as in Eq. (1) based on each of these training sets, and evaluate the emotion classification results with these learned classifiers respectively on the test set. Finally, the classification results are stacked together to demonstrate the improvement of text emotion classification with increasing number of samples in the training sets, as shown in Fig. 2.

As the training set grows with the selected samples by Algorithm I, the learned logistic regression emotion classification model is incrementally improved with steadily increasing micro Precision, Recall, and F1 scores for text emotion classification. Specifically, as the active learning procedure is carried out for 60 loops, the micro average scores of Precision, Recall, and F1 have been improved by 7.53%, 7.36%, and 7.51%, respectively. The results suggest that our approach is effective in finding the appropriate samples from raw data set to significantly improve the learning of a multi-label text emotion classification model.

Next, we examine the effectiveness of ours propose complementariness criterion by comparing the text emotion classification results with the results from a control experiment in which the sample selection procedure consists of only the first three criteria, i.e. the informativeness, the representativeness, and the diverseness. The selection parameters in the control experiment are kept the same as before, except the second argument of argpartition for U^D is replaced by λ^C to keep the size of sample selection results the same as Algorithm I.

In Fig. 2, the micro average scores of Precision, Recall, and F1 from the proposed approach are consistently higher than those from the control experiment. The averaged gaps between two approaches are 1.55% in Precision, 0.94% in Recall, and 1.30% in F1 scores. The comparison of two experiments indicates that the proposed complementariness criterion is able to reorder the priority for sample selection in an effective manner so that high-quality samples can be more easily found and added into the training set. When the comparison comes to CIRD model (proposed by Kang et al. [28]), increments of the micro average scores are 1.55%, 2.49% and 1.97% for Precision, Recall and F1, respectively.

Finally, we look into the distribution of emotion labels generated by the proposed approach and by the control experiment, to further analyze the label-balancing property in these algorithms. Figure 3 shows the growing number of emotion labels in the training set as increasing number of samples are selected by the proposed active learning algorithm (Fig. 3a) or the controlling algorithm (Fig. 3b) and are annotated with the emotion tags by human experts. The proposed active learning algorithm

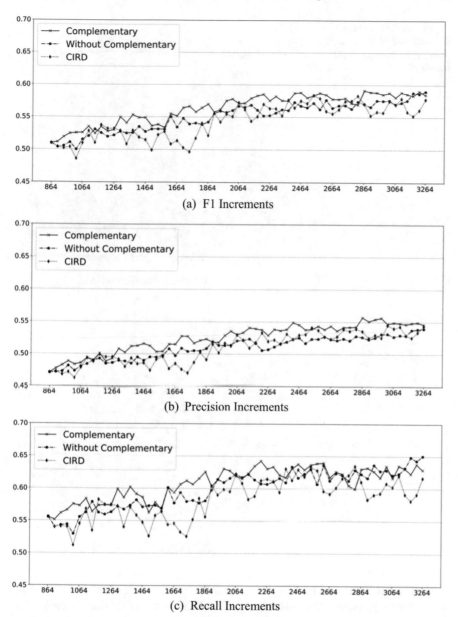

(a) F1 Increments

(b) Precision Increments

(c) Recall Increments

Fig. 2 Increments of emotion classification with respect to the incrementally increased training data

Fig. 3 Increment of each
kind of emotion labels in the
training set through active
learning

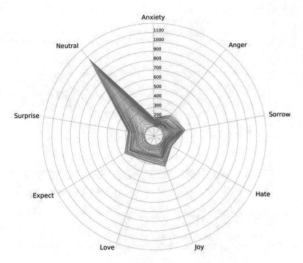

(a) with balance criterion

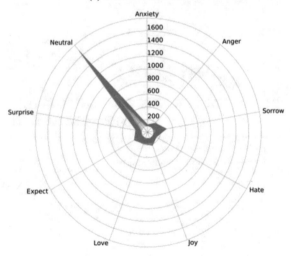

(b) without balance criterion

has generated a series of training sets with more balanced emotion labels than those
generated by the controlling algorithm. Specifically, the growth of the neutral emotion
label, which turns to be the most frequent label in the raw data sets, has been significantly restrained in the selection procedure. Meanwhile, the other emotion labels
are growing much faster than those selected by the controlling algorithm, and the
growth is especially significant for Anxiety, Joy, Hate, and Expect. These results
suggest that given raw data sets of highly biased label distributions, the proposed
complementariness criterion is effective in selecting raw samples of label-balancing
property, which essentially restrains the number of samples with frequent labels and

increases the number of samples with infrequent labels in the training set, through active learning.

5 Conclusion

In this paper, we proposed a novel complementariness criterion through active learning to keep the balance of emotion label distributions in the training set of continuously growing number of samples, which have been selected from the raw data sets with a potentially highly-biased emotion label distribution. The complementariness criterion was designed to restrain the selection of samples with potentially frequent labels and to discover those with potentially infrequent labels, through an explicit evaluation of the Kullback–Leibler divergence of the emotion label distributions in the temporal training sets, which was composed of the existing training set and a candidate raw sample for selection, to the ideal uniform emotion distribution. Experiment results suggested that the proposed criterion was effective in balancing the label distribution in the generated training sets for emotion classification, and that the text emotion classification results were steadily improved with the growing of label-balanced training data. Our future work would be focused on the improvement of optimizing the model to achieve better performance on generating the examples to make the label's distribution more balanced.

Acknowledgements This research has been partially supported by the Ministry of Education, Science, Sports and Culture of Japan, Grant-in-Aid for Scientific Research(A), 15H01712.

References

1. Lippmann W (2017) Public opinion Routledge
2. Ravi K, Ravi V (2015) A survey on opinion mining and sentiment analysis. tasks, approaches and applications. Knowl-Based Syst 89:14–46
3. González-Bailón S, Paltoglou G (2015) Signals of public opinion in online communication: a comparison of methods and data sources. ANNALS Am Acad Polit Soc Sci 659(1):95–107
4. López-de-Ipiña K, Alonso JB, Travieso CM et al (2013) On the selection of non-invasive methods based on speech analysis oriented to automatic Alzheimer disease diagnosis. Sensors 13(5):6730–6745
5. Friedman HS, Booth-Kewley S (1987) Personality, type A behavior, and coronary heart disease: the role of emotional expression. J Pers Soc Psychol 53(4):783
6. Tacconi D, Mayora O, Lukowicz P et al (2008) Activity and emotion recognition to support early diagnosis of psychiatric diseases. In: Pervasive computing technologies for healthcare. In: Second international conference on pervasive health 2008. IEEE, pp 100–102
7. Yu LC, Wu JL, Chang PC et al (2013) Using a contextual entropy model to expand emotion words and their intensity for the sentiment classification of stock market news. Knowl-Based Syst 41:89–97
8. Rao Y, Xie H, Li J et al (2016) Social emotion classification of short text via topic-level maximum entropy model. Inf Manag 53(8):978–986

9. Das S, Chen M (2001) Yahoo! for amazon: extracting market sentiment from stock message boards. In: Proceedings of the Asia Pacific finance association annual conference, vol 35
10. Desmet P (2003) A multilayered model of product emotions. Des J 6(2):4–13
11. Dave K, Lawrence S, Pennock DM (2003) Mining the peanut gallery: opinion extraction and semantic classification of product reviews. In: Proceedings of the 12th international conference on World Wide Web. ACM, pp 519–528
12. Hassenzahl M, Diefenbach S, Göritz A (2010) Needs, affect, and interactive products–facets of user experience. Interact Comput 22(5):353–362
13. Wang L, Ren F, Miao D (2016) Multi-label emotion recognition of weblog sentence based on Bayesian networks. IEEJ Trans Electr Electron Eng 11(2):178–184
14. Picard RW (1995) Affective computing
15. Ekman P (1992) An argument for basic emotions. Cogn Emot 6(3–4):169–200
16. Zhao YY, Qin B, Liu T (2010) Sentiment analysis. J Softw 21(8):1834–1848
17. Taboada M, Brooke J, Tofiloski M et al (2011) Lexicon-based methods for sentiment analysis. Comput Linguist 37(2):267–307
18. Yang C, Lin KHY, Chen HH (2007) Building emotion lexicon from weblog corpora. In: Proceedings of the 45th annual meeting of the ACL on interactive poster and demonstration sessions. Association for Computational Linguistics, pp 133–136
19. Matsumoto K, Ren F (2011) Estimation of word emotions based on part of speech and positional information. Comput Hum Behav 27(5):1553–1564
20. Picard RW, Vyzas E, Healey J (2001) Toward machine emotional intelligence: analysis of affective physiological state. IEEE Trans Pattern Anal Mach Intell 23(10):1175–1191
21. Bhowmick PK (2009) Reader perspective emotion analysis in text through ensemble based multi-label classification framework]. Comput Inf Sci 2(4):64
22. Tsoumakas G, Katakis I (2007) Multi-label classification: an overview. Int J Data Warehouse Min (IJDWM) 3(3):1–13
23. Tsoumakas G, Vlahavas I (2007) Random k-labelsets: an ensemble method for multilabel classification. In: European conference on machine learning. Springer, Berlin, Heidelberg, pp 406–417
24. Ren F, Kang X (2013) Employing hierarchical Bayesian networks in simple and complex emotion topic analysis. Comput Speech Lang 27(4):943–968
25. Liu H, Lieberman H, Selker T (2003) A model of textual affect sensing using real-world knowledge. In: Proceedings of the 8th international conference on Intelligent user interfaces. ACM, pp 125–132
26. Reyes O, Morell C, Ventura S (2018) Effective active learning strategy for multi-label learning. Neurocomputing 273:494–508
27. Li X, Guo Y (2013) Active learning with multi-label SVM classification. IJCAI, 1479–1485
28. Kang X, Wu Y, Ren F (2018) Progressively improving supervised emotion classification through active learning. In: International conference on multi-disciplinary trends in artificial intelligence. Springer, Cham, pp 49–57

Arbitrary Perspective Crowd Counting via Local to Global Algorithm

Chuanrui Hu, Kai Cheng, Yixiang Xie, and Teng Li

Abstract Crowd counting is getting more and more attention. More and more collective activities, such as the Olympics Games and the World Expo, are also important to control the crowd number. In this paper, we address the problem of crowd counting in the crowded scene. Our model accurately estimated the count of people in the crowded scene. Firstly, we proposed a novel and simple convolutional neural network, called Global Counting CNN (GCCNN). The GCCNN can learn a mapping, transforms the appearance of image patches to estimated density maps. Secondly, The Local to Global counting CNN (LGCCNN), calculating the density map from local to global. Stiching the local patches constrains the final density map of the larger area, which make up for the difference values in the perspective map. In general, it makes the final density map more accurate. The dataset we used is a set of public dataset, which are WorldExpo'10 dataset, Shanghaitech dataset, the UCF_CC_50 dataset and the UCSD dataset. The experiments have proved our method achieves the state-of-the-art result over other algorithms.

Keywords Crowd density map · Convolutional neural network · Perspective distortion

1 Introduction

The crowd counting has important social significance and market value. Managers can reasonable scheduling of manpower, material resources and optimize resources configuration by using the number of ROI area statistics. For some of the square, passageway and other public occasions, the result of crowd statistics have very good warning effect to social security problems. Therefore, the crowd counting becomes

---

Chuanrui Hu and Kai Cheng: These authors contributed equally to this work.

C. Hu · K. Cheng · Y. Xie · T. Li (✉)
Anhui University, NO. 111 Jiulong RD, Hefei 230061, China
e-mail: liteng@ahu.edu.cn

© Springer Nature Switzerland AG 2021
H. Lu (ed.), *Artificial Intelligence and Robotics*,
Studies in Computational Intelligence 917,
https://doi.org/10.1007/978-3-030-56178-9_3

Fig. 1 Sample crowd scene from the WorldExpo'10 dataset

the key point in the field of video analysis and intelligent video surveillance. This involves estimating the number of people in the crowd and the crowd distribution over the entire region.

Traditional crowd counting algorithms share a common procedure. (1) Foreground segmentation, but the split of foreground can not entirely separate people and background because sometimes people are still in high density scenarios, such as the queue in front of the station ticket window. (2) Crowd feature extraction, due to the dense scenarios perspective distortion, brightness condition and low resolution of the image, the handcraft features (e.g., Scale Invariant Feature Transform (SIFT) [13], Histogram of Oriented Gradient (HOG) [3], Local Binary Patterns (LBP) [17]) cannot fully express characteristics of the crowd. For the dense crowd, typical static crowd scenes from the WorldExpo'10 Dataset [2].

It is difficult to detect the number of people because of occlusion, and it is not wise to calculate the number by the foreground segmentation due to the randomness of foreground segmentation. Some typical scenes from the WorldExpo'10 dataset [2] are shown in Fig. 1.

There has been a significant recent progress in the field of crow counting due to the successful development of deep learning (e.g., the convolutional neural networks (ConvNets)) [2, 4, 12, 14, 15]. To the best of our knowledge, the Cong et al. [2]

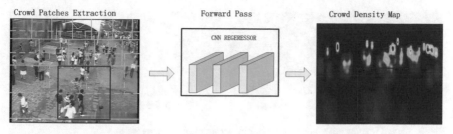

Crowd Patches Extraction   Forward Pass   Crowd Density Map

CNN REGERESSOR

Fig. 2 We define the crowd counting task like a regression problem where a CNN model to map the appearance of image to crowd density map. The yellow box indicates that the training image dataset is densely extracted from the whole image

first train a CNN model to learn a map to solve the crowd counting problem, but in order to get the crowd count, the result need to feed a ridge regressor with the output features. The MCNN [21], which output is an estimated density map and it solve the large scale variation. But the output final estimated density map is distortion due to the size of the the final estimated density map is decreased. Recent researches [2, 21] have proven the learned features performed better than the traditional hand-crafted features. As illustrated in Fig. 2, in order to make up for the shortcoming of the resent search [2, 21], we propose our convolutional neural network architectures to learn the regression function that mapping the image appearance into a crowd density map. The number of people in the crowd scene is calculated through integration over the crowd density map

The main contributions of this work can be conclude into these three aspects.

- In Sect. 3.1, we propose a novel convolutional neural network architectures, named Global Counting CNN (GCCNN). Which is a fully convolutional neural network [12] can get an accurate regression of a crowd density map of image patches. We adopted a bilinear interpolation algorithm, the Fig. 4 clearly shows that the final output feature map is the same size as the input patch.
- Due to the scale variation in the crowd images, we introduce the Local to Global Counting CNN (LGCCNN) in Sect. 3.2 which provide an algorithm, calculating the final density map form local to global. The algorithm make up for the differences caused by different values in the perspective map then makes the density map of the larger area more accurate.
- Our architecture has been evaluated on three benchmark datasets and is shown to achieve state-of-the-art outperforms.

The rest of this paper are organized as follows: previous research about the crowd counting is in Sect. 2. The proposed method and the overall structure of the two CNN model are detailed listed in Sect. 3. Experiments and the comparisons of results are summarized in Sect. 4. In the end, we make a conclusion about this paper in Sect. 5.

| Target Image | Groundtruth | Target Image | Groundtruth |

Fig. 3 Crowd images with their corresponding groundtruth density maps

2 Related Works

In recent years, the crowd counting method in the literature can be divided into two categories: counting by detection and counting by regression.

Counting by detection [5, 10, 16, 20]. Many algorithms counting people by detection. First, they use the appearance and the motion feature to separate the moving objects from the background over the two consecutive frames of a video clip. Then these algorithms utilize the handcraft features (such as Haar wavelet features or edgelet features [20]) to obtain the moving objects. However these methods can be used in the video clip not suit for the still image and the handcrafted features often sustain a decline in accuracy when the scene is perspective distortion, severe overlapping, and varying illumination.

Counting by regression [1, 2, 7–9, 11, 14, 19, 21]. Counting by regression aims to lean a mapping between the low-level features and people count via certain a regression function without foreground segmentation or pedestrian detection. It is more suitable for complex environments and more crowded instance like pedestrians. Cong et al. [2] first trained a deep CNN model. It makes good performance. But they reported the results feeding a ridge regressor with the output features of their CNN model and the input patches of their CNN model is random which does not consider the large scale-invariant to large scale changes well. Our network diminish the perspective distortion and estimates both the crowd count as well as the crowd density map.

3 Methodology

In this section, we will state our notation and crowd counting methodology. Here, we treated the crowd counting problem as the density map estimation.

Previous research has followed [2] and defined the groundtruth of the density map regression as sum of the Gaussian kernels centered on the locations of objects. This mentioned density map is more suitable for representing the density distribution of circle-like objects like cells and bacteria. Considering the shape of the pedestrian in an ordinary surveillance camera is ellipse-like. We follow the method of the [2].

Before generate the groundtruth density map, we should consider the large scale variation due to the perspective distortion. Perspective normalization is necessary to describe the pedestrian scale. After we get the perspective map of each scene and a set of head annotations images, where all the heads are marked by dots. We can generate the groundtruth density map D_i, for an image I, is defined as a sum of Gaussian functions centered on each dot annotation. We generate the crowd density map is generated as:

$$D_i(p) = \sum_{P \in P_i} \frac{1}{\|Z\|} (\mathcal{N}_h(P; P_h, \sigma_h) + \mathcal{N}_b(P; P_b, \Sigma)) \tag{1}$$

where P_i is the set of 2D points of the image I, $\mathcal{N}_h(P; P_h, \sigma_h)$ and $\mathcal{N}_b(P; P_b, \Sigma)$ respectively represent a normalized 2D Gaussian kernel as a head part and a normalized 2D Gaussian kernel as a body part. P_h is the head position and P_b is the body position, estimated by the head position value and the value in the perspective map. Some groundtruth density maps is shown in Fig. 3. Our CNN model is to learn a non-linear regression function that takes an image patch P with associated groundtruth density map and groundtruth crowd count. As an assistant object, the crowd count associated with the training patch is integrated from groundtruth density map. It returns an estimated density map $D_{pred}^{(P)}$.

$$D_{pred}^{(P)} = F(P|\theta_{net}) \tag{2}$$

where θ_{net} is the set of parameters of the CNN model. For the image patch P, we could get the $D_{pred}^{(P)}$. Thus for a given unseen test image, at first our algorithm densely extracted image patches over the image. Then our CNN model could generate an estimated density map corresponding to the image patch. At last, all the density maps are aggregated into a density map for a whole test image.

3.1 The Global Counting CNN Model

Let us introduce our first ConvNet structure called the Global Counting CNN Model (GCNN). As illustrated in Fig. 4. The crowd density estimation does not like image classification, it need per pixel predictions. So we adopt the fully convolutional neural networks natural. This would reduce the overfitting due to the fully convolutional neural network has much fewer parameters than a network trained on an entire image. The structure consists of 6 convolution layers and 2 pooling layers. They are specially designed to extract the crowd features. The Conv1 layer has 3×3 filters with a depth of 64. The Conv2 layer has 3×3 filters with a depth of 128. The max pooling layer with a 2×2 kernel size is used after conv1 and conv2. The Conv1 layer has 33 filters with a depth of 256. The Conv4 layer and Conv5 layer are made of 1×1 filters with a depth of 1000 and 400. The Conv6 layer is another 1×1 filters with a depth of 1.

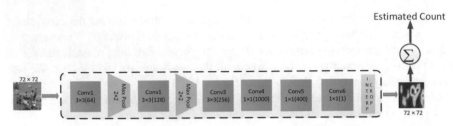

Fig. 4 Our GCCNN structure, treated the input patches and their associated groundtruth density maps and groundtruth crowd counts as input, which returns an estimated density map, the size is same as the input patch

The output from these convolution layers is upsampled to the size of the input image patch using bilinear interpolation to directly obtain the estimated crowd density map.

Due to the good performance for the CNNs of the Parametric Rectified Linear Unit (PReLU) [6]. The PReLU was adopted as the activation function and it is not shown in the Fig. 2. Equation (2) has point out, our CNN models is to learn a mapping from a set of features extracted from training image patches to an estimated crowd density map. So, our GCCNN is trained to solve the regression problem. The Euclidean distance is used as the loss function.

$$L1(\theta net) = \frac{1}{2N} \sum_i^N ||F(Pi|\theta net) - D_{gt}^{Pi}||^2 \qquad (3)$$

$$L2(\theta net) = \frac{1}{2N} \sum_i^N ||C(Pi|\theta net) - C_{gt}^{(Pi)}||^2 \qquad (4)$$

where θnet denotes the learned parameters of the CNN model, N is the number of the training images, Pi is the image patch will be training in the CNN model. $F(Pi|\theta net)$ and $C(Pi|\theta net)$ represent the corresponding image patch stand for the estimated crowd density map and the crowd count. D_{gt}^{Pi} and C_{gt}^{pi} respectively represent the groundtruth density map and ground truth crowd number of the corresponding image patch. Different from Zhang et al., the master loss task is the $L1(\theta net)$. We let the two loss functions pass through all previous layers together. The master loss task is the $L1(\theta net)$. The $L2(\theta net)$ is treated as the auxiliary loss. The auxiliary loss task helps optimize the learning process, while the master loss task takes the most responsibility. We add weight to balance the auxiliary loss. The two loss tasks assisted each other and trained together to obtain optimization.

After obtaining the parameters θnet of the CNN model. How do we implement the prediction stage on the unseen target test image? First, we densely extracted image patches. Then all the image patches are resized to 72×72 pixels. These input image patches with their associated groundtruth density maps and groundtruth crowd count are as illustrated in Fig. 5, which passed through our CNN architecture. It returns an

Groundtruth Count=2. 01

Groundtruth Count=6. 37

Groundtruth Count=6. 19

Groundtruth Count=3. 51

Fig. 5 Patches with their associated labels

estimated density map corresponding to the input image patch. Lastly, all the output estimated density map will be aggregated into a density map over the whole test image. Due to the extracted image patches are overlap. So the each location of the final estimation density map must by normalized according the number of patches that calculated into the final estimated density map.

3.2 The Local to Global Counting CNN Model

On the basis of a counting by regression model, using the annotated perspective map of each scene to solve the perspective distortion and scale variation. Due to the impact of the perspective distortion on each image, the size of pedestrian will exhibit scale variation. The features extracted from the same pedestrian at different scene depths would have notable differences in values.

Go a step further, in order to get an accurate estimated crowd density map, we use the Local to Global Counting CNN Model (LGCCNN). We proposed an algorithm for estimating a density map from local to global which is specialized in the perspective distortion and scale variation. The ConvNet structure was specialized designed is shown in Fig. 6. Our CNN model was consisted of three columns CNN. The three parallel CNNs contain the same structure (i.e., conv-pooling-conv-pooling) and the same size of filters. The CNN model takes different but related inputs. The input is the training image patches cropped from the training images. The patch of the first column was resized to 94×94 pixel. The next two columns take the upper and lower two parts of a complete patch. Each parallel CNN is in charge of learning features of input patch for a different perspective value. Then the output feature maps of the last two columns CNN Model are stitched. Compared the losses of the GCCNN, we added the same loss function as show in Eq. (7).

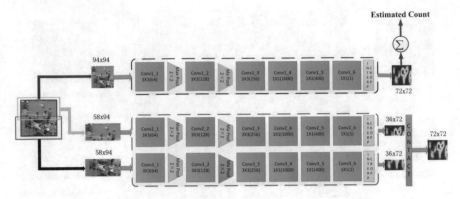

Fig. 6 The CNN architecture for LGCCNN

$$L1\,(\theta net) = \frac{1}{2N} \sum_i^N ||F\,(Pi|\theta net) - D_{gt}^{Pi}||^2 \qquad (5)$$

$$L2\,(\theta net) = \frac{1}{2N} \sum_i^N ||C\,(Pi|\theta net) - C_{gt}^{(Pi)}||^2 \qquad (6)$$

$$L3\,(\theta net) = \frac{1}{2N} \sum_i^N ||F\,(Pi1;\,Pi2|\theta net) - F(Pi|\theta net)||^2 \qquad (7)$$

where θnet denotes the learned parameters of the LGCNN model, N is the number of the training images, P_i is the image patch will be training in the CNN model. P_{i_1} and P_{i_2} are the corresponding upper and lower image patch of the completed image patch P_i. $F(Pi|\theta net)$ is the output feature map of the first column CNNmodel. Noticed, the upper and lower estimated density maps were calculated by the different values of the perspective map. We constrain the final estimated density map on the lager region which makes up for the difference caused by the different perspective values. In the end, it makes the estimated density map more accurate and provide a method for crowd density map was from local to global.

4 Experiments

We first evaluate our CNN model on the challenging the WordExpo'10 dataset [2]. The detail of the WorldExpo'10 dataset is shown in Table 1. This dataset contains 1132 annotated video clips, captured by 108 surveillance cameras. 1,127 one-minute long video sequences are treated as training datasets. Testing datasets, 5 one-hour long different video sequences. Each video sequence contains 120 labeled frames.

Table 1 The attribution of the public datasets: NUM is number of frames; Total is the number of labeled people; MAX is the maximum number of people in the ROI of a frame; MIN is the minimum number of people in the ROI of a frame. AVG indicated the average crowd count

Dataset	NUM	Total	MAX	MIN	AVG
UCSD	2000	49885	46	11	25
UCF_CC_50	50	63974	1279	4543	1279
WorldExpo'10	4.44 million	199623	253	1	50
ShanghaiTech Part A	482	241677	3139	33	501
ShanghaiTech Part B	716	88488	578	9	123

We train our deep convolution neural network on the basis of caffe library and some modifications are applied. The NVIDIA GTX TITAN X GPU is used. We use the standard Stochastic Gradient Descent (SGD) algorithm to optimize ConvNet parameters with a learning rate of 1e−7 and momentum of 0.9 video clips captured by 108 surveillance cameras. 1,127 one-minute long video sequences are treated as training datasets. Testing datasets, 5 one-hour long different video sequences. Each video sequence contains 120 labeled frames. We train our deep convolution neural network on the basis of caffe library and some modifications are applied. The NVIDIA GTX TITAN X GPU is used. We use the standard Stochastic Gradient Descent (SGD) algorithm to optimize ConvNet parameters with a learning rate of 1e−7 and momentum of 0.9.

4.1 Evaluation Criteria

In order to make the experimental results more intuitionistic. We use the two evaluation criteria: the mean absolutely evaluation (MAE) and the mean square evaluation (MSE) which are defined as follows:

$$MAE = \frac{1}{N} \sum_{1}^{N} |Ci - Ei| \tag{8}$$

$$MSE = \sqrt{\frac{1}{N} \sum_{1}^{N} |Ci - Ei|} \tag{9}$$

where N denotes the number of the training images, C_i is the true pedestrians number of the ith test image. E_i is the estimated pedestrians number of ith test image. MAE represents the actual situation of the estimates error. MSE represent the robustness of the estimates. The lower MAE and MSE are, the count result is more accurate and better.

4.2 Data Preprocessing

The dataset consists of 108 scenes. In order to train our GCCNN model, we typically selected 2600 images from the 103 scenes in the dataset. We collected 200 patches of 72×72 pixels extracted all over the image with their associated groundtruth density maps and groundtruth crowd counts. It contains 100 positive patches which center is the area of people and the 100 negative patches which center is the area of the ground. Then we performed a data augmentation by flipping each patch randomly.

To train our LGCCNN model, we performed the same method as previously mentioned to extract the image patches of 94×94 pixels. For a given complete patch of 94×94 pixels, we will get the upper and lower patch of 58×94 pixels by cropped the complete patch.

We typically selected 2600 images from the 103 scenes in the dataset as the training images. Firstly, we collected 200 patches of 94×94 pixels extracted all over the image with their associated groundtruth density maps and groundtruth crowd counts. It contains 100 positive patches which center is the area of people and the 100 negative patches which center is the area of the ground. Then we perform a data augmentation by flipping each patch randomly. After we got these patches and in order to meet our CNN model. We will collect the small local patches by cropped the global patches.

4.3 Results

Table 2 shows a comparison of our models with the state-of-the-art approaches on the WorldExpo'10 dataset. The first (Rodriguez et al. [18]) use LBP+RR are traditional algorithm. The Zhang et al. and MCNN are CNN-based algorithm. Table 3 shows the results with other 3 public dataset. The datasets prepocessing is like the WorldExpo'10 dataset.

Table 2 Quantitative results with other state of the art methods on the WorldExpo'10 dataset

Method	Scene 1	Scene 2	Scene 3	Scene 4	Scene 5	Avg
LBP+RR [18]	13.6	58.9	37.1	21.8	23.4	31.0
Cong et al. [2]	9.8	14.1	14.3	22.2	**3.7**	12.9
MCNN [21]	3.4	20.6	12.9	13.0	8.1	11.6
GCCNN	7.5	22.6	15.7	16.0	6.2	13.6
LGCCNN	**2.6**	**19.3**	**17.4**	**14.8**	4.7	**11.0**

Table 3 Comparing performances of different methods on Shanghaitech dataset, the UCF_CC_50 dataset and the UCSD dataset

| Method | The Shanghaitech dataset | | | | The UCF_CC_50 | | The UCSD | |
| | Part A | | Part B | | Dataset | | Dataset | |
	MAE	MSE	MAE	MSE	MAE	MSE	MAE	MSE
LBP+RR [18]	303.2	371.0	59.0	81.7	–	–	–	–
Cong et al. [2]	181.8	277.7	32.0	49.8	467.0	**498.5**	1.60	3.31
MCNN [21]	110.2	173.2	26.4	41.3	337.6	509.1	1.07	1.35
LGCCNN	**105.2**	**169.8**	**25.6**	**40.3**	**336.5**	510.2	**1.05**	**1.27**

We started to analyze the experiment result. Our GCCNN data preprocessing like the method of Zhang et al. [4]. In the Table 2, our GCCNN model get a better performance in scene 1 and scene 5. These two subdatasets contains pedestrians is about 80 in each image which is more suitable in the actual world. By contrast, the best performance is LGCCNN, which is reduced the MAE effectively. The LGCCNN combined the local and the global features, is well to solve the problem of perspective distortion and scale variation. Figure 7 shows some of qualitative results of the WorldExpo'10 dataset that are obtained by the LGCCNN model.

5 Conclusions

In this paper, we proposed two convolution neural network architectures. For our first architecture, the GCCNN model can learn a mapping which transforms the appearance of crowd image to the crowd density map effectively. Our second architecture, the LGCCNN model which goes a step further, provide a method for crowd density map was from local to global. The final estimated density map on the lager region which makes up for the difference caused by the upper and lower image patch of different perspective values. In the end, it makes the estimated density map more accurate. The density map is generated in the output layer of network and the number of people is obtained by integral regression. We test our proposed method in the Shanghaitech dataset, the WorldExpo'10 dataset, the UCF_CC_50 dataset and the UCSD dataset. Moreover, the experimental results show the accuracy the robustness of our method outperforms the state-of-the-art crowd counting method.

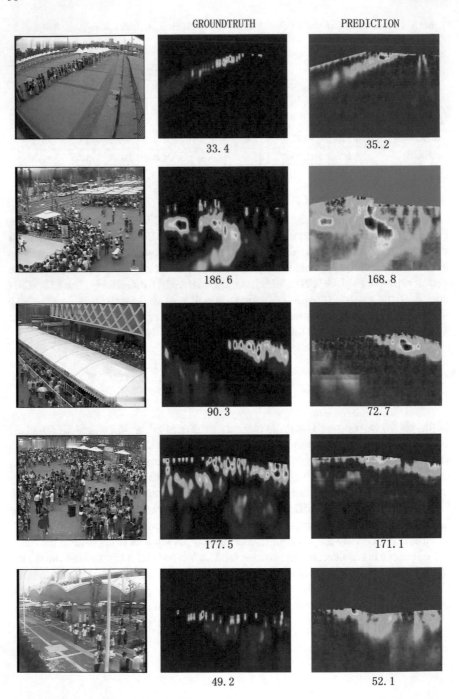

Fig. 7 Sample predictions of our LGCCNN model in the WorldExpo'10 dataset. The first column is the target test image. The second column is the groundtruth density map corresponding to the target test image. The second column shows the estimated density map

Acknowledgements This work is supported by the National Natural Science Foundation (NSF) of China (No. 61572029), and the Anhui Provincial Natural Science Foundation of China (No. 1908085J25), and Open fund for Discipline Construction, Institute of Physical Science and Information Technology, Anhui University.

References

1. Chan AB, Vasconcelos N (2012) Counting people with low-level features and bayesian regression. IEEE Trans Image Process 21(4):2160–2177
2. Cong Z, Li H, Wang X, Yang X (2015) Cross-scene crowd counting via deep convolutional neural networks. In: IEEE conference on computer vision & pattern recognition
3. Dalal N, Triggs B (2005) Histograms of oriented gradients for human detection. In: Schmid C, Soatto S, Tomasi C (eds) International conference on computer vision & pattern recognition (CVPR '05), vol 1. IEEE Computer Society, San Diego, United States, pp 886–893. https://doi.org/10.1109/CVPR.2005.177, https://hal.inria.fr/inria-00548512
4. Ge S, Zhao S, Li C, Li J (2018) Low-resolution face recognition in the wild via selective knowledge distillation. CoRR. http://arxiv.org/abs/1811.09998
5. Ge W, Collins RT (2009) Marked point processes for crowd counting. In: IEEE conference on computer vision & pattern recognition
6. He K, Zhang X, Ren S, Jian S (2015) Delving deep into rectifiers: surpassing human-level performance on imagenet classification
7. Hu Y, Chang H, Nian F, Yan W, Teng L (2016) Dense crowd counting from still images with convolutional neural networks. J Visual Comm Image Represent 38(C):530–539
8. Ke C, Chen CL, Gong S, Tao X (2012) Feature mining for localised crowd counting. In: British machine vision conference
9. Lempitsky VS, Zisserman A (2010) Learning to count objects in images. In: International conference on neural information processing systems
10. Lin SF, Chen JY, Chao HX (2001) Estimation of number of people in crowded scenes using perspective transformation. IEEE Trans Syst Man Cybernetics Part A Syst Humans 31(6):645–654
11. Liu T, Tao D (2014) On the robustness and generalization of cauchy regression. In: IEEE international conference on information science & technology
12. Long J, Shelhamer E, Darrell T (2015) Fully convolutional networks for semantic segmentation. In: Proceedings of the IEEE conference on computer vision and pattern recognition, pp 3431–3440
13. Lowe DG (1999) Object recognition from local scale-invariant features. In: Proceedings of iccv
14. Loy CC, Gong S, Xiang T (2014) From semi-supervised to transfer counting of crowds. In: IEEE international conference on computer vision
15. Lu H, Wang D, Li Y, Li J, Li X, Kim H, Serikawa S, Humar I (2019) Conet: a cognitive ocean network. CoRR. http://arxiv.org/abs/1901.06253
16. Min L, Zhang Z, Huang K, Tan T (2009) Estimating the number of people in crowded scenes by mid based foreground segmentation and head-shoulder detection. In: International conference on pattern recognition
17. Ojala T, Pietikinen M, Menp T (2002) Gray scale and rotation invariant texture classification with local binary patterns. IEEE Trans Pattern Anal Machine Intell 24(7):971–987
18. Rodriguez M, Laptev I, Sivic J, Audibert JY (2011) Density-aware person detection and tracking in crowds. In: International conference on computer vision
19. Sam DB, Surya S, Babu RV (2017) Switching convolutional neural network for crowd counting

20. Wu B, Nevatia R (2005) Detection of multiple, partially occluded humans in a single image by bayesian combination of edgelet part detectors. In: Tenth IEEE international conference on computer vision
21. Zhang Y, Zhou D, Chen S, Gao S, Yi M (2016) Single-image crowd counting via multi-column convolutional neural network. In: Computer vision & pattern recognition

A Semi-supervised Learning Method for Automatic Nuclei Segmentation Using Generative Adversarial Networks

Chuanrui Hu, Kai Cheng, Jianhuo Shen, Jianfei Liu, and Teng Li

Abstract Nuclei segmentation is a fundamental step to measure cell motility and morphology in High-throughput microscopy images. One prerequisite of automatic nuclei segmentation is to manually label a huge amount of cell images, which is a labor-intensive and time-consuming process. This paper develops a semi-supervised learning approach to reduce the dependence on the amount of labeled images, and it consists of three main steps. First, cell regions are automatically extracted from the background through a novel convolutional neural network (CNN). Comparing with the state-of-art segmentation network structure, our designed network structure has fewer parameters and combines layers of the feature hierarchy that refines the spatial precision of the output probability map. Second, discriminator networks are explored to produce confidence maps for unlabeled microscopy images by applying a fully convolution network (FCN) as the discriminator network. It also can be used to improve the performance of the segmentation network by coupling the adversarial loss with standard cross-entropy loss. Finally, combining both labeled images and unlabeled images with confidence maps to train the segmentation network can identify individual nuclei on the extracted cell regions. Experimental results on only 84 images showed that SEG score could achieve 77.9% and F1 score could achieve 76.0%, which outperformed existing using 670 images segmentation methods. Such promising result suggested that the proposed semi-supervised nuclei segmentation approach could achieve high segmentation accuracy with a small set of labeled images, which is a desirable property to analyze microscopy images in the real clinical practice.

Chuanrui Hu and Kai Cheng: These authors contributed equally to this work.

C. Hu · K. Cheng · J. Shen · J. Liu · T. Li (✉)
Anhui University, NO. 111 Jiulong RD, Hefei 230061, China
e-mail: liteng@ahu.edu.cn

© Springer Nature Switzerland AG 2021
H. Lu (ed.), *Artificial Intelligence and Robotics*,
Studies in Computational Intelligence 917,
https://doi.org/10.1007/978-3-030-56178-9_4

Keywords Cell segmentation · Semi-supervised learning · Generative adversarial networks

1 Introduction

Recent development of personal medicine requires the treatment should be tailored to the individual with the best response and highest safety margin. The key to personal medicine is to quantitatively assess individual biology, which is often acquired from cell morphology and tissue structure through microscopy image analysis. Quantitative analysis of microscopy image involves the measurements of cell density, shape, size, texture, and other imagenomics. Manual measurement fails to meet the clinical needs of processing microscopy images in a high-throughput rate because it is a lab-intensive and time-consuming process. Manual measurement is also irreproducible, which makes it difficult to be used in longitudinally tracking disease progression.

Automatic nuclei segmentation is of particular interest because it can produce repeatable measurements. However, there often exist poor contrast between cell/nuclei regions and background, some nuclei in microscopy images are shown in Fig. 1.

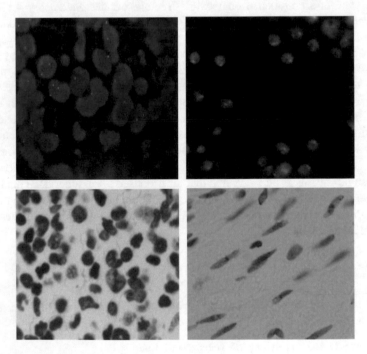

Fig. 1 Some examples of nuclei image from microscopy

Potentially caused by image noise and cell motility during image acquisition, nucleus shapes, sizes, and texture often exhibit significant variance. Nucleus touching is also present in the image regions with densely-packing cells. Many methods based on characteristic of microscopy images have been made to address these issues, such as gray level thresholding, watershed [19] and active contour [2, 11]. Although mathematical models are deliberately designed to fit image characteristics in these methods, they are prone to being violated due to the complexity of microscopy images. Deep learning based methods, on the other hand, O. Ronneberger et al. proposed U-Net [15] structure and Ba et al. [8] proposed deep multiple instance learning to segment microscopy images, they can build mathematical models directly from microscopy images. Accurate nuclei segmentation can be achieved by utilizing a large number of labelled images. Unfortunately, labelling microscopy images is an extremely tedious process, which is sometimes infeasible in the real clinical practice.

Labeling microscopy images is an extremely tedious process which motivates us to develop a semi-supervised learning strategy to reduce the dependence on the number of labelled images. The success of generative adversarial networks (GANs) hints us to understand such nuclei similarity by producing confidence maps for unlabeled images to increase the training data. Based on this idea, we developed a novel semi-supervised learning-based nuclei segmentation method, which can be summarized in three fold.

- A light neural network structure is developed to identify nuclei in the microscopy images, which is called Light-Unet. A multi-layer up-sampling network can enhance edge information specialized in our nuclei image segmentation. Experiments demonstrated that Light-Unet performed faster than existing networks, such as U-Net.
- An adversarial framework is designed to improve nuclei segmentation accuracy without additional computation in the inference process. We adopt the adversarial learning through adversarial loss in discriminator network. With this adversarial loss, we maxmize the probability of the predicted segmentation map being considered as the ground truth distribution.
- For unlabeled images, we utilized semi-supervised learning scheme leveraging on the discriminator network to assist the segmentation network training well.

The rest of this paper consists of the following sections: We review the deep learning methods for cell image segmentation and semi-supervised learning for cell image segmentation in Sect. 2; The overall framework and the proposed method are detailed in Sect. 3; Experiments and the comparisons of results are summarized in Sect. 4; Finally, we conclude this paper in Sect. 5.

2   Related Works

In this section, we first review the related work of deep learning methods for biomedical image segmentation in recent years, and then briefly summarize the related semi-supervised methods for cell image segmentation.

2.1   Deep Learning Methods for Cell Image Segmentation

Cell segmentation is an important task in biomedical image analysis and attracts lots of research efforts in recent years. Recent state of the art methods for sematic segmentation are mostly based on CNN. Long et al. [9] proposed a fully convolutional neural network (FCN). Pixel classification level based CNN is first involved in segmentation task. But it is not insensitive to details in the image, that means that detailed spatial information is obliterated by the repeated combination of max-pooling and downsampling performed at every layer of standard convolutional neural networks (CNNs). Noh et al. [12] proposed a multi-layer up-sampling network structure which effectively enhanced edge information. Motivated by the their successes in regular image semantic segmentation, deep learning based methods have become popular for cell segmentation such as: [8, 14, 15], and they have shown more promising results than traditional segmentation algorithms. Raza et al. [14] proposed the MIMO-Net to solve the problem of segmentation. Ge et al. [5] use the knowledge distillation to decompress the deep neural networks. O. Ronneberger et al. proposed U-Net for biomedical image segmentation. U-Net is a structure which skips connections from the encoder features to the corresponding decoder activations. The network can localize to capture object details. And there are also many others cell segmentation networks based on U-Net.

2.2   Semi-supervised Learning for Cell Image Segmentation

For the task of cell image segmentation, pixel-level annotation is usually expensive and requires professional knowledge. To reduce the heavy effort of labeling segmentation ground truth, in recent years semi-supervised learning and weakly-supervised methods have been explored for cell image segmentation such as [1, 17, 20, 21].

In the weakly-supervised setting, the segmentation network can be trained with bounding boxes level labeling (Dai et al. [4]; Khoreva et al. [7]), without the need for pixel level labeled data. However these weakly supervised approaches still required a large amount of labeled data, while the detailed boundary information is difficult to capture with bounding box labels. Semi-supervised methods can leverage the unlabeled images for model learning. Su et al. [18] propose an interactive cell segmentation method by classifying feature-homogeneous super-pixels into specific classes, which is guided by human interventions. Xu et al. [21] propose a CNN with a semi-supervised regularization to address the neuron segmentation in 3D volume with introducing a regularization term to the loss function of the CNN such that the performance is improved.

Recently, the generative adversarial networks (GANs) originally proposed by Goodfellow et al. [6] have shown to be quite useful and the idea is also applied in semi-supervised image segmentation. Luc et al. [10] proposed about semantic segmentation using adversarial network, and the output of their discriminator is a vector rather than a map. Souly et al. [16] proposed a method for semi-supervised semantic segmentation using GANs, but these generated examples may not be sufficiently close to real images to help the segmentation network.

In this paper, we propose a novel deep learning based semi-supervised method for cell image segmentation. Our proposed semi-supervised algorithm based on GANs considers the output of discriminator as the supervisory signals, and learn the confidence map through the discriminator network as the instructor for semi-supervised learning.

3 Methodology

In this section, we first provide an overview of our proposed algorithm, then we introduce the designed segmentation network structure and full convolutional discriminator network structure, finally we introduce the proposed detailed learning scheme of segmentation network and discriminator network.

3.1 Algorithm Overview

Figure 2 shows the overview of our proposed method. Our method has two networks, a discriminator network training and a segmentation network training. We denote the segmentation network as S(·) and the discriminator network as D(·). Given an input image with size of H × W × 3. The segmentation network outputs probably map of size H × W × C. where C is equal to 2 in our work. The discriminator network outputs a spatial probability map of size H × W × 1.

For the first network: a discriminator network. A typical discriminator network is to distinguish ground truth label maps and probability maps from segmentation network. Our discriminator network is inspired by fully convolutional network (FCN) [9]. The input is either from the segmentation network output segmentation map $S(X_n)$ or from the ground truth label, the ground truth label need to be one hot encode before training, and then outputs spatial probability map. Each pixel value in the discriminator network output maps represent the probability that pixel is from the segmentation map or ground truth label. In contrast to typical GANs, which takes a fixed size image as input and outputs is a single probability value that reflect a image

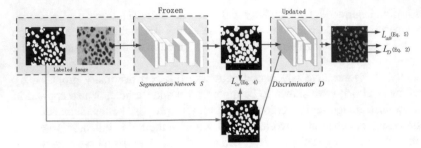

a. Update discriminator D using labeled image

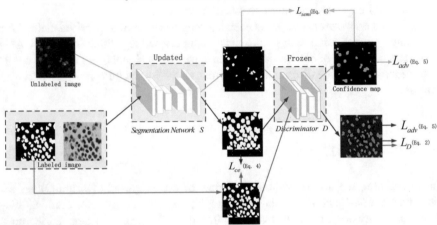

b. Update segmentation network S using both labeled image and unlabeled image

Fig. 2 A detailed description of our proposed algorithm. We optimize the fully convolutional discriminator with loss L_D. We optimize the segmentation network by using three loss function:cross-entropy loss L_{ce}, adversarial loss L_{adv} and semi-supervised loss L_{semi}

is real or fake. Our discriminator network is a fully-convolutional network can take a arbitrary size image, and output a probability map. It reflects a spatial level. Notice that discriminator network's parameters is only updataed with the labeled images training.

For the second network: a nuclei segmentation network. Given a input image with size H × W × 3, output the class probability map of size H × W × 2. When training with labeled images, we train the segmentation network by optimizing standard cross-entropy loss (L_{ce}) with ground truth label and adversarial loss (L_{adv}) with discriminator network. Noticed that, we only use labeled images to train the discriminator network. When training with unlabeled images, we train the segmentation map with adversarial loss with discriminator network and semi-supervised loss. Noticed that our segmentation network's parameters is updated both with labeled images and unlabeled images.

During the training process, we use both label and unlabeled images under the semi-supervised setting. When training with labeled images, we train the segmentation network by optimizing standard cross-entropy loss (L_{ce}) with ground truth label and adversarial loss (L_{adv}) with discriminator network. For the unlabeled data, we train the segmentation network with adversarial loss with discriminator network and semi-supervised loss. We obtain the confidence map by the initial segmentation map $S(X_n)$ passing through discriminator network. The confidence map like an instructor to teach the segmentation network to select confidential region in probability map $S(X_n)$. We then perform a semi-supervised loss with confidential regions. Note that the segmentation network is always supervised by adversarial loss due to the adversarial loss is only related to the discriminator network.

3.2 Network Architecture

3.2.1 Segmentation Network

Our segmentation network structure is inspired by U-Net. We designed network contains 6 convolution layers, 2 upppooling layers. Also batch-normal and Xavier's initialization is perform after every layer. Due to our network has much fewer parameter than U-Net. We named it as light-unet. Similar with U-Net, its upsampling part we have a large number of feature channels, which allow the network to propagate context information to higher resolution layers. The structure is shown in Fig. 3 and the parameters of each layer are listed in Table 1:

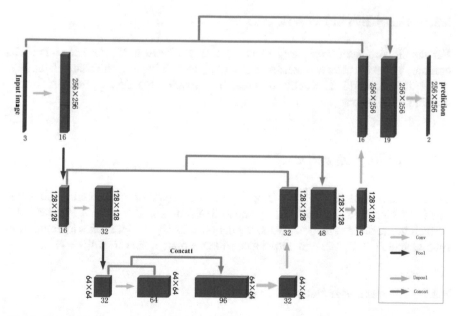

Fig. 3 The structure of our proposed Light-Unet. Each box corresponds to a multi-channel feature map. The number of channels is denoted on bottom of box. Our Light-Unet fusion multi layer information to refine the spatial precision of the output

Table 1 The relevant layers and their parameters of Light-Unet

Layer	Layer type	Parameters
Conv1	Convolution	Filter size: 5 × 5, Filter number: 16, padding: 2
Pool1	Pooling	Pooling method: Max, Kernel size: 2 × 2
Conv2	Convolution	Filter size: 3 × 3, Filter number: 32, padding: 1
Pool2	Pooling	Pooling method: Max, Kernel size: 2 × 2
Conv3	Convolution	Filter size: 5 × 5, Filter number: 64, padding: 1
Concat1	Concation	Pool2, Conv3
Conv4	Deconv	Filter size: 1 × 1, filter number: 32, padding: 1
Uppool1	Maxunpool	Filter size: 2 × 2
Concat2	Concation	Pool1, Deconv1
Conv5	Convolution	Filter size: 1 × 1, filter number: 16, padding: 1
Uppool2	Maxunpool	Kernel size: 2 × 2
Concat3	Concation	Input, Conv5
Conv6	Convolution	Filter size: 1 × 1, filter number: 2, padding: 1

3.2.2 Discriminator Network

For the discriminator work, we follow the structure used in Radford et al. [13]. It contains 5 convolutions with kernel size is 4×4 with stride is 2 with channel number is $\{64, 128, 256, 512, 1\}$. Leaky-ReLU, is the activation function applied after every convolutional layer.

3.3 Training Objective

Given an input image X_n of size $H \times W \times 3$, we denote the segmentation network as S($\cdot$) which predicted probability map as $S(X_n)$ of size $H \times W \times 2$. For our fully convolutional discriminator network, we denote it as D($\cdot$) which takes a probability map of size $H \times W \times 2$ and output a confidence map of size $H \times W \times 1$.

3.3.1 Discriminator Network Training

In order to train this network, according to Goodfellow et al. [6], they employed an adversarial loss, the function is defined as:

$$\min_G \max_D V(D, G) = E_{x \sim P_{data}(x)}[log D(x)]$$
$$+ E_{z \sim P_z(z)}[log(1 - D(G(z)))] \tag{1}$$

where x represent the original image from an unknown distribution P_{data}, z is the input noise of the generator network $G(\cdot)$, and $G(z)$ is the output map of generator network. $D(\cdot)$ and $G(\cdot)$ are playing in a min-max game with this loss function. In other word, we can transform equation (1) into another form, that likes:

$$L_D(\theta_{dis}) = - \sum_{h,w}((1 - y_n)log(1 - D(S(X_n)|\theta_{dis}))^{(h,w)} +$$
$$y_n log(D(Y_n|\theta_{dis})^{(h,w)})) \tag{2}$$

where θ_{dis} is the discriminator network parameters. X_n is the n-th input image. When $y_n = 0$, It means the input of the discriminator network is the predicted segmentation map of segmentation network. When $y_n = 1$, It means the input image is sampled from the ground truth label. Notice that, ground truth label image is only one channel, we need to perform one-hot decode (convert the ground truth channel the same as the probability map $S(X_n)$) before training.

3.3.2 Segmentation Network Training

In this stage, we propose a multi-task loss to optimize the segmentation network. The loss can be defined as:

$$L_{seg}(\theta_{seg}) = L_{ce}(\theta_{seg}) + \lambda_{adv}L_{adv}(\theta_{seg}) + \lambda_{semi}L_{semi}(\theta_{seg}) \tag{3}$$

where θ_{seg} is the discriminator network parameters. L_{seg}, L_{adv}, L_{semi} represents the cross-entropy loss, adversarial loss, and the semi-supervised loss respectively. There are two hyper-parameter λ_{adv}, λ_{semi}. The role of these parameters are used to balance the multi-task training.

Training with labeled image. When we train with labeled image, we can use the stand cross-entropy loss to optimize it. The loss is defined as:

$$L_{ce}(\theta_{seg}) = - \sum_{h,w} \sum_{c \in C} Y_n^{(h,w,2)} log(S(X_n|\theta_{seg})^{(h,w,2)}) \tag{4}$$

where $S(X_n)$ is the prediction result of the segmentation network for the n-th input image.

With standard cross-entropy loss L_{ce} we can make the predict probability map distribution more close to the ground truth label distribution.

We adopt adversarial learning by using adversarial loss, the loss is defined as:

$$L_{adv}(\theta_{seg}) = -\sum_{h,w} log(D(S(X_n|\theta_{seg}))^{(h,w)}) \tag{5}$$

With adversarial loss L_{adv}, we can maximum the probability of the predicted segmentation map $S(X_n)$ be considering of the ground truth distribution.

Training with unlabeled image. When training with unlabeled image. Obviously, L_{ce} will not applied into multi-task. L_{adv} is still available, since it only rely on discriminator network. The discriminator network can generate spatial probability map. The confidence map reflects that probability that the prediction results are close to the distribution of the ground truth map. We then binarized confidence map by set a threshold T_{semi}. Final, we denote a ground truth label as the masked segmentation prediction $\hat{Y}_n = argmax(S(X_n))$ combining this binarized confidence map. The semi-supervised loss is define as:

$$L_{semi}(\theta_{seg}) = -\sum_{h,w}\sum_{c\in 2} I(D(S(X_n))^{(h,w)} > T_{semi})$$
$$\cdot \hat{Y}_n^{(h,w,2)} log(S(X_n|\theta_{seg})^{(h,w)}) \tag{6}$$

where $I(\cdot)$ is the active function, and T_{semi} is the threshold to control the confidential regions. During training, we have already get the $\hat{Y}_n^{(h,w,2)}$, we can just consider it as a standard cross-entropy loss.

With utilize the semi-supervised loss L_{semi}, we can enhance the segmentation network's performance by using more unlabeled images.

Our goal is to learn the parameters θ_{dis} in the discriminator network and θ_{seg} in the segmentation network, while θ_{dis} is parameters in the discriminator network only introduced propagated the labeled image training and the parameters are updated by Adaptive Moment Estimation (Adam). θ_{seg} is parameters in the segmentation network. The parameters θ_{seg} are updated by Stochastic Gradient Descent (SGD).

4   Experiments

4.1   Implementation Details

The proposed algorithm is implement on the basis of pytorch library and some modifications are applied. The NVIDIA GTX TITAN X GPU is used. The standard Stochastic Gradient Descent (SGD) algorithm is applied to optimize segmentation network parameters with the momentum of 0.9, batch size of 10 and weight decay of 0.0004. The initial learning rate is set as 0.001. For training the discriminator network, we adopt Adam optimizer with learning rate as 0.0001 with momentum set as 0.999.

For semi-supervised training phase, we randomly selected the labeled images and unlabeled image in every iteration. Before start semi-supervised learning, we start training only with labeled images in top 600 iterations in our experiments. Due to the model will suffer from initial noisy masks and predictions and after 600 iterations the initial segmentation map of segmentation network is good. In each iteration, when training with labeled images, we update the segmentation network and discriminator network jointly. when training with unlabeled images, we only update the segmentation network utilized semi-supervised learning.

After we get the experiments result, we conduct watershed in opencv library which are useful for sticky nuclei.

4.2 Experiment Setup

4.2.1 Dataset

Our dataset is collected from 2018 Data Science Bowl competition. This dataset contains a large number of segmented nuclei images. The images were acquired under a variety of conditions and vary in the cell type, magnification, and imaging modality (brightfield vs. fluorescence). There are 3 types images. Some images are shown in Fig. 4: The images are of different sizes and so preprocessing is required to make them of uniform size (256, 256). There are 670 images in the training set and 65 images in the test set.

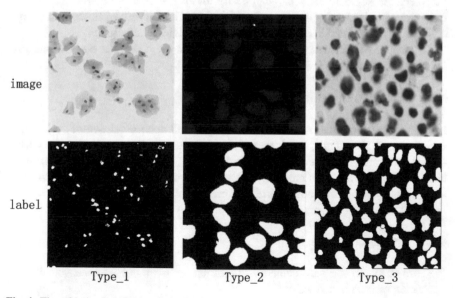

Fig. 4 Three kinds of nuclei image in the dataset

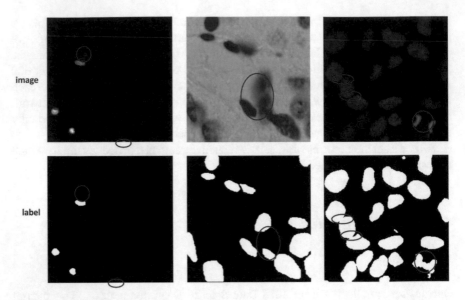

Fig. 5 Some weakly annotation nuclei images

Some nuclei images might annotated with some errors. The wrong picture is shown in Fig. 5. As can been from the Fig. 5, some nuclei images may have some questions: The first and second column image shows there have a obvious nucleus in image, but the manual segmentation may ignore it. In the test phase, we find that our predicted segmentation map can easily find it however the human expert may ignore it. The third column image shows some individual nuclei in the microscope cell image, but manual annotation cause some nuclei may clustered.

In the end, we selected 40 images in the original test dataset as our test images.

4.2.2 Data Augmentation

It has been proven that small dataset or complicated network structure may lead to serious over-fitting. Thus, in this paper, we adopt two data augmentation categories. One is horizontal rotation Another is random crop. We first rescale images with the size of 331×331 and then randomly select crops with the size of 256×256. Adding augmentated data makes our algorithm more robustness.

4.2.3 Evaluation Criteria

F-Score (F1): To evaluate detection performance, we use F1 to measure it. The evaluation criteria is defined as:

$$F1 = \frac{2 \cdot P \cdot R}{P + R} \tag{7}$$

where P and R are the precision ($P = \frac{TP}{TP+FP}$) and recall ($R = \frac{TP}{TP+FN}$). True positives (TP) are nuclei with intersection over union overlap (IOU) > 0.5 with an unmatched ground truth (GT) nuclei segmentation masks, otherwise it is counted as false positive (FP). GT nuclei which remain unmatched are counted as false negative (FN). The higher F1 score means our algorithm can identify more nuclei in image.

SEG: We evaluate accuracy of predicted segmentation masks using the SEG measure, based on Jaccard similarity index. Jaccard similarity index is computed using $J(G, S) = \frac{|S \cap G|}{|S \cup G|}$. The SEG can be denote as:

$$SEG = \frac{\sum_{i=1}^{n} \sum_{j=0}^{m} J(S_{ij} \cap G_{ij})}{n \cdot m} \tag{8}$$

where i is i-th input nucleus test image, j is j-th nucleus in the test image. SEG is mean of Jaccard similarity index of all GT nuclei and ranges between 0 to 1. The higher SEG means the more similar between segmentated nuclei images and ground truth label images. The higher SEG score means the individual nucleus is more similar than original manual annotation label.

4.3 Comparison with the State-of-the-Art Methods

First we report the experiments results of our proposed methods and comparisons with other deep learning methods. The Table 2 shown it. Chen et al. [3] applied Deeplab-v2 network structure. we use it as segmentation network to identify cell's nuclei in microscopy images. Ronneberger et al. [15] proposed U-Net to identify cells successfully. The Table 2 shows our model is get a better result than DeepLab-v2 and a little precious loss compare with U-Net. But our model has much fewer parameters than U-Net. Our Light-Unet is under the setting about $L_{adv} = 0.002$, L_{semi} = 0.1, T_{semi}=0.2. The detailed comparison with U-Net is shown in Table 3.

In order to validate the semi-supervised scheme, we randomly select 1/32, 1/16, 1/8, images as labeled and used the rest of the training data as unlabeled. By using adversarial loss L_{adv}, the model achieved about 1.4–1.9% over the baseline model in F1 score. This means our adversarial loss scheme can boost our segmentation network learn the structural information from ground truth label distribution. Then when both use adversarial loss L_{adv} and semi-supervised loss L_{semi}, the model can increase about 2.8–3.2% F1 score again than our baseline model.

Sometimes, we find that when we apply Deep-v2 structure, and select 1/8 amount of dataset, we can get a best SEG score 78.1%. Does it means this algorithm is more precious than another method ? The segmentation result is need to combine both SEG and F1 score to make a accurate decision. It can get a highest SEG score, however it's F1 score is only 65.6%. It means many nuclei in image can not complete identify.

Table 2 Comparison with other deep learning methods

Methods	Data amount							
	1/32		1/16		1/8		Full	
	SEG	F1	SEG	F1	SEG	F1	SEG	F1
DeepLab-v2	74.3	55.5	76.5	60.3	**78.1**	65.6	77.7	67.0
U-Net	74.5	62.1	77.0	70.0	77.8	75.2	78.1	**75.5**
Light-Unet (our baesline)	73.9	60.7	76.0	68.8	78.0	73.3	78.1	73.3
Light-Unet + L_{adv}	73.4	62.4	76.1	70.2	77.9	74.9	**78.1**	75.2
Light-Unet + L_{adv} + L_{semi}	**74.5**	**64.7**	**76.1**	**72.0**	77.9	**76.0**	78.0	**76.1**

Table 3 Comparison between U-Net and light-unet

Network structure	Parameter size	Inference time (ms)
U-Net	132M	0.078
Light-Unet	236k	0.029

In addition, we find the highest F1 score is merely 76.1%. It suggests approximately one fourth nuclei are not clearly identified. Sometimes a frame can contains more than 300 nuclei, in this situation, even human experts can not identify most nuclei clearly. The nuclei number of each test image is range from a dozen to a few hundred. In contrast to most existing cells or nuclei dataset, which contains the number of cells or nuclei about 30 each frame. When testing with small number of nuclei in image, our proposed method can achieve at least 86.0%.

4.4 Hyper-parameters Analysis

The segmentation task adopts multi-task learning strategy to learn the two hyper parameters λ_{adv} and λ_{semi}. We used T_{semi} to control the sensitivity in the semi-supervised learning. We show the results with different parameter setting in Table 4. We first evaluate the effect on λ_{adv}. Note that without λ_{adv} loss, the model achieves 75.5% F1 score and 78.1% SEG score. When λ_{adv} set equal to 0.002, the model will achieve 1.2% improvement in F1 score and 0.2% improvement in SEG score. When $\lambda_{adv} = 0.005$, the model performance will descend about 1.5% in F1 score. It means adversarial loss is too large.

Second, we show comparisons under different value of λ_{semi} with λ_{adv} equal to 0.002 and select 1/16 amount of data. We set λ_{semi} in {0.05, 0.1, 0.2}, Only with L_{adv} loss, the baseline model get a 70.2% F1 score and 76.1% SEG score. In general, when $L_{semi} = 0.1$, the model will achieve a best performance with a 1.8% F1 score improvement than without L_{semi} loss.

Table 4 Hyper parameters analysis

Data amount	λ_{adv}	λ_{semi}	T_{semi}	SEG	F1
Full	0	0	–	78.1	75.5
Full	0.001	0	–	**78.4**	75.9
Full	0.002	0	–	78.3	**76.7**
Full	0.003	0	–	78.1	76.1
Full	0.004	0	–	78.2	75.4
Full	0.005	0	–	78.3	74.0
1/16	0.002	0	–	76.1	70.2
1/16	0.002	0.05	0.2	75.9	71.1
1/16	0.002	0.1	0.2	76.0	**72.0**
1/16	0.002	0.2	0.2	**76.2**	69.9
1/16	0.002	0.1	0	75.8	69.5
1/16	0.002	0.1	0.1	76.0	71.2
1/16	0.002	0.1	0.2	**76.1**	**72.0**
1/16	0.002	0.1	0.3	75.7	71.5
1/16	0.002	0.1	0.4	75.9	70.4
1/16	0.002	0.1	1	76.1	70.2

Lastly, we perform experiments with different value of T_{semi}, where we set $\lambda_{adv} = 0.002$, $\lambda_{semi} = 0.1$ and T_{semi} in $\{0.1, 0.2, 0.3, 0.4, 0.5\}$. For unlabeld images, the higher T_{semi}, we only select ground label map which has a structural similarity more close to the ground truth label distribution. When we set $T_{semi} = 0$, it means, we trust all the pixel in the predicted segmentation map of unlabeled image, the F1 score is drop to 69.5%. The other extreme, the T_{semi} is set to equal to 1. It means we don't truth any region of the predicted result of predicted segmentation map. The algorithm perform best, when we set $T_{semi} = 0.2$. It achieves 72.0% F1 score and 75.9% SEG score. The nuclei segmentation results are shown in Fig. 6. We train the light-unet without adversarial loss L_{adv} and semi-supervised loss L_{semi} as our baseline model.

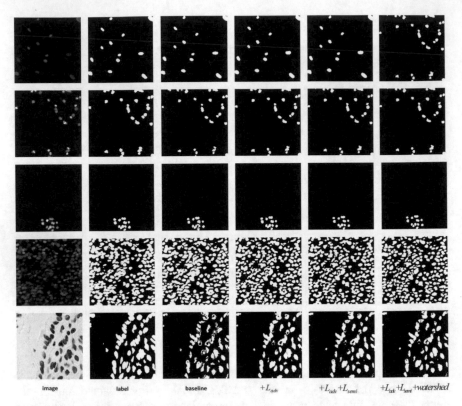

image   label   baseline $+L_{adv}$ $+L_{adv}+L_{semi}$ $+L_{adv}+L_{semi}+watershed$

Fig. 6 Our nuclei segmentation result using 1/16 training data. The first row show the nuclei are well separated. we called the original image as 'easy' image. The third row has many cluster nuclei, we called image like this as 'hard' image. For 'hard' image, we find that the mannal annotation image even can not split individual each nucleus well, Sometimes, the predicted segmentation image can split cluster nucleus more effective than manual segmentation image. Take a deep step, we perform watershed on segmentation result images

5   Conclusions

In this paper, we developed a semi-supervised learning method to automatically segment nuclei regions on microscopy images. Inspired by generative adversarial networks, our method consists of a segmentation network and a discriminator network. The segmentation network is initialized through a small set of labelled microscopy images. It is continuously improved by treating unlabeled images with confidence maps from the discriminator network as the training data. The loss function of the segmentation network is also updated with an adversarial loss. Our method achieved high segmentation accuracy on a small set of microscopy images, which is comparable with existing methods on a larger number of labeled images.

Our semi-supervised learning strategy has a great potential in real clinical usages. Most nuclei segmentation methods required a large number of labelled images to support their supervised-learning scheme despite the fact that manual labeling by experts is both costly and time-consuming. On the contrary, our semi-supervised method only needs a small set of representative labelled images, which is more acceptable by experts to establish training data. Moreover, our method is not limited to a particular nucleus type, and it can be used as a general segmentation framework on different types of microscopy images.

The primary reason to cause segmentation errors is the over-segmentation of touching nuclei, which often happened in densely-clustered nuclei regions. We are currently trying to either embed cell edge confidence maps during training process [15] or watershed in post-processing to separate nuclei clusters. In the future, we also need to determine the minimum number of labelled image to achieve clinically-acceptable segmentation accuracy. Nevertheless, our semi-supervised learning method could achieve high segmentation accuracy using a small set of labelled images, illustrating the potential for our method as a clinical tool for high-throughput microscopy image analysis.

Acknowledgements This work is supported by the National Natural Science Foundation (NSF) of China (No. 61702001), and the Anhui Provincial Natural Science Foundation of China (No. 1908085J25), and Open fund for Discipline Construction, Institute of Physical Science and Information Technology, Anhui University.

References

1. Arbelle A, Raviv TR (2018) Microscopy cell segmentation via adversarial neural networks. In: 2018 IEEE 15th international symposium on biomedical imaging (ISBI 2018). IEEE, pp 645–648
2. Bamford P, Lovell B (1998) Unsupervised cell nucleus segmentation with active contours. Signal Process 71(2):203–213
3. Chen LC, Papandreou G, Kokkinos I, Murphy K, Yuille AL (2018) Deeplab: semantic image segmentation with deep convolutional nets, atrous convolution, and fully connected crfs. IEEE Trans Pattern Anal Mach Intell 40(4):834–848
4. Dai J, He K, Sun J (2015) Boxsup: exploiting bounding boxes to supervise convolutional networks for semantic segmentation. In: Proceedings of the IEEE international conference on computer vision, pp 1635–1643 (2015)
5. Ge S, Zhao S, Li C, Li J (2018) Low-resolution face recognition in the wild via selective knowledge distillation. CoRR abs/1811.09998. http://arxiv.org/abs/1811.09998
6. Goodfellow I, Pouget-Abadie J, Mirza M, Xu B, Warde-Farley D, Ozair S, Courville A, Bengio Y (2014) Generative adversarial nets. In: Advances in neural information processing systems, pp 2672–2680
7. Khoreva A, Benenson R, Hosang J, Hein M, Schiele B (2017) Simple does it: weakly supervised instance and semantic segmentation. In: Proceedings of the IEEE conference on computer vision and pattern recognition, pp 876–885
8. Kraus OZ, Ba JL, Frey BJ (2016) Classifying and segmenting microscopy images with deep multiple instance learning. Bioinformatics 32(12):i52–i59
9. Long J, Shelhamer E, Darrell T (2015) Fully convolutional networks for semantic segmentation. In: Proceedings of the IEEE conference on computer vision and pattern recognition, pp 3431–3440

10. Luc P, Couprie C, Chintala S, Verbeek J (2016) Semantic segmentation using adversarial networks. arXiv:1611.08408
11. Meijering E, Dzyubachyk O, Smal I (2012) Methods for cell and particle tracking. In: Methods in enzymology, vol 504. Elsevier, pp 183–200
12. Noh H, Hong S, Han B (2015) Learning deconvolution network for semantic segmentation. In: Proceedings of the IEEE international conference on computer vision, pp 1520–1528
13. Radford A, Metz L, Chintala S (2015) Unsupervised representation learning with deep convolutional generative adversarial networks. arXiv:1511.06434
14. Raza SEA, Cheung L, Epstein D, Pelengaris S, Khan M, Rajpoot NM (2017) Mimo-net: a multi-input multi-output convolutional neural network for cell segmentation in fluorescence microscopy images. In: 2017 IEEE 14th international symposium on biomedical imaging (ISBI 2017). IEEE, pp 337–340
15. Ronneberger O, Fischer P, Brox T (2015) U-net: convolutional networks for biomedical image segmentation. In: International conference on medical image computing and computer-assisted intervention. Springer, pp 234–241
16. Souly N, Spampinato C, Shah M (2017) Semi and weakly supervised semantic segmentation using generative adversarial network. arXiv:1703.09695
17. Su H, Yin Z, Huh S, Kanade T (2013) Cell segmentation in phase contrast microscopy images via semi-supervised classification over optics-related features. Med Image Anal 17(7):746–765
18. Su H, Yin Z, Huh S, Kanade T, Zhu J (2016) Interactive cell segmentation based on active and semi-supervised learning. IEEE Trans Med Imaging 35(3):762–777
19. Vincent L, Soille P (1991) Watersheds in digital spaces: an efficient algorithm based on immersion simulations. IEEE Trans Pattern Anal Mach Intell 6:583–598
20. Wei Y, Liang X, Chen Y, Shen X, Cheng MM, Feng J, Zhao Y, Yan S (2017) STC: a simple to complex framework for weakly-supervised semantic segmentation. IEEE Trans Pattern Anal Mach Intell 39(11):2314–2320
21. Xu K, Su H, Zhu J, Guan JS, Zhang B (2016) Neuron segmentation based on cnn with semi-supervised regularization. In: Proceedings of the IEEE conference on computer vision and pattern recognition workshops, pp 20–28

Research on Image Encryption Based on Wavelet Transform Integrating with 2D Logistic

Xi Yan and Xiaobing Peng

Abstract In order to protect them from theft, it is necessary to encrypt image data. When considering the security algorithm of image information security, we must consider its particularity. Two-dimensional Logistic chaotic mapping is introduced in the design of pixel gray value substitution, and two substitution ideas are adopted. That is to say, the low-frequency coefficient matrix after wavelet decomposition is scrambled first, and the image after wavelet transform is replaced by the pixel value. The encryption system constructed by the chaotic dynamic model can still solve the information better after the multimedia information is processed by some signals. The combination of the lifting wavelet transform in the frequency domain and the chaotic sequence in the spatial domain is used for image encryption, which has good encryption effect and practical prospect.

Keywords Digital image · Logistic integration · Wavelet transform · Image encryption

1 Introduction

Image information is one of the most important means for human to express information. The exchange and transmission of image data through the public network is simple, fast and not subject to geographical restrictions, which can save a lot of costs for data owners [1]. With the increasing frequency of information exchange and the rapid increase of storage capacity, the analysis of image data with huge storage capacity has become one of the main means for people to obtain information [2]. Not all images can be disclosed, and some of them need to be kept secret in

X. Yan (✉)
Department of Computer, Jiangsu University of Science and Technology, Zhenjiang 212003, China
e-mail: just_yx@qq.com

X. Peng
School of Computer Science and Communication Engineering, Jiangsu University, Zhenjiang 212003, China

© Springer Nature Switzerland AG 2021
H. Lu (ed.), *Artificial Intelligence and Robotics*,
Studies in Computational Intelligence 917,
https://doi.org/10.1007/978-3-030-56178-9_5

the process of receiving. For example, military facility drawings, weapons drawings and satellite pictures [3]. In the process of image transmission, it is often attacked artificially. Including information theft, data tampering, virus attacks, etc. [4]. Image information security has a very broad meaning. When considering its security algorithm, its particularity must be considered. Data owners need reliable image data encryption technology to protect their own interests [5]. Traditional secrecy focuses on restricting access to data, and data expansion is not considered when designing algorithms [6]. The chaotic dynamic system has pseudo-random type, certainty and extreme sensitivity to initial conditions and system parameters. Therefore, it can construct a very good information encryption system [7]. How to ensure the information security of images has become a hot topic of extensive attention and research by experts and scholars.

Image encryption is actually a disorder of the pixels of an image in accordance with a written algorithm or a replacement of pixel values. The information of the original image is hidden, and what the interceptor sees during the sending and receiving process is an unordered picture [8]. The party receiving the image can restore the image according to the algorithm's key. In the field of image encryption, emerging security algorithms are constantly being proposed and researched [9]. The encryption system constructed by the chaotic dynamic model can still solve the information better after the multimedia information is processed by some signals [10]. Before chaotic encryption, image data is transformed by wavelet transform. If one of the wavelet coefficients is changed, it will be reflected in all pixel points by inverse operation of wavelet transform. The problem of image information security has a very broad meaning. When considering its security algorithm, we must consider its particularity [11]. The encryption system based on chaotic dynamic model can still declassify the information [12] well after the image information is processed by some signal processing. Chaotic sequence generated by two-dimensional Logistic mapping is used to adjust the template of wavelet transform coefficients and chaotic scrambling method, which can obtain encrypted images with high security.

2   Materials and Methods

The encryption algorithm based on the hybrid theory is widely used because it fully considers the characteristics of the image itself. Cryptocoding mainly studies the method of transforming an information to protect it from being stolen, interpreted and utilized by the enemy in the process of transmission of the channel. Because of the high sensitivity and randomness of chaotic map to initial conditions, chaotic sequences generated can be used to encrypt and decrypt information. The complete cryptosystem also includes the sender and the receiver, the cryptographic algorithm and the attacker of the password. The encryption system constructed by chaotic dynamic model can solve the multimedia information well after it is processed by some signals [13]. If the attacker does not crack the key and decrypts the image data stream directly, the decryption process cannot be achieved at all. After the

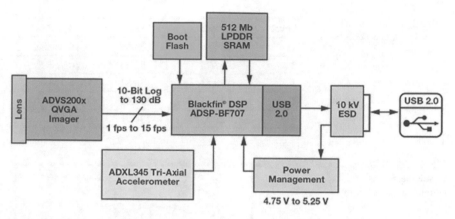

Fig. 1 Digital image analysis system structure

image information is decomposed by the two-dimensional wavelet decomposition algorithm, a series of sub-images with different resolutions can be obtained. From a spatial distribution, an image can be thought of as a rectangular block of several pixels [14]. Figure 1 shows the structure of a digital image analysis system.

In the process of transmitting information, it is very likely that an eavesdropper will intercept the ciphertext information. In practical applications, due to the limitation of computer word length, the use of low-dimensional chaotic sequences will degenerate into periodic sequences, which is vulnerable to the chaotic model reconstruction method. The low-frequency coefficients and high-frequency coefficients of the two-dimensional signal are extracted from the decomposition structure, and chaotic cat mapping is performed on the low-frequency coefficients and the high-frequency coefficients respectively, thus completing the image encryption. Each algorithm has its own limitations, such as being related to periodicity. In order to obtain the decrypted image, first the coefficients are inversely mapped by cat mapping, and then the extracted low-frequency coefficients and high-frequency coefficients are used for two-dimensional wavelet reconstruction. The image diffusion is fused, in particular, the scrambled pixel values are modified to achieve the purpose of image diffusion. After the fusion of scrambling and diffusion, a good encryption effect can be obtained through several iterations (Table 1).

The performance parameters of image encryption before and after optimization are shown in Table 2. After wavelet transform and filtering optimization, the Logistic

Table 1 Performance parameters of digital image structure before and after optimization		Before optimization	After optimization
	Rows	136	112
	Number of columns	97	65
	Monitoring points	346	534

Table 2 Evaluation of denoising effects at different scales

Level	PSNR	MSE	Time
1	28.36	98.53	0.869
2	27.75	95.36	0.758
3	30.51	146.81	1.123
4	31.68	112.63	1.117
5	28.73	136.57	1.545

Fig. 2 Digital image wavelet optimization simulation comparison

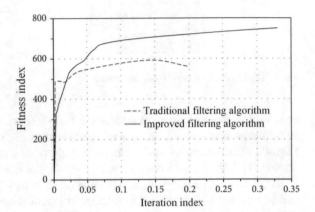

integration structure has been greatly optimized, with fewer nodes and better monitoring area. Logistic integrated reliability optimization simulation is compared with Fig. 2.

The multi-scale product of the larger amplitude wavelet coefficients tends to be larger, and the multi-scale product of the smaller amplitude wavelet coefficients tends to be smaller. For wavelet decomposition of one-dimensional signals, the one-dimensional multi-scale product at point t is:

$$\omega_j = d_j / \sum_{j=1}^{p} d_j = (1 - e_j) / \left(p - \sum_{j=1}^{p} e_j \right) \tag{1}$$

The two-dimensional multiscale product is defined as:

$$I_i = \left[\sum_{j=1}^{p} \omega_j^m y_{ij}^m \right]^{1/m} \tag{2}$$

Let i_t be the natural image to be processed and do a wavelet transform on i_t:

$$i_t = r^* + \pi_t + \alpha(\pi_t - \pi^*) + \beta \tilde{y}_t \tag{3}$$

Password analysis is to recover a message without knowing the key. A successful cryptanalyst can recover the plaintext or key of the message. After the mixed sequence is modulo, it is converted into a bit sequence of the corresponding number of bits. Similarly, the scrambled pixel values are converted into bit sequences of corresponding numbers of bits, and then the two are XORed bit by bit. The wavelet decomposition scale will greatly affect the effect of image processing. If the characteristics of wavelet coefficients under multi-scale can be fully utilized, the quality of reconstructed images can be improved to some extent. The denoising effects obtained at different scales are shown in Table 2.

Image encryption based on chaotic cat mapping is to redistribute the pixel points of the original image through cat mapping, making the original image messy [15]. Since the method takes into account the influence of operations such as compression of image data on the encrypted data stream, it conforms to the popular image coding scheme. Therefore, it has strong adaptability and broad application prospects.

3 Result Analysis and Discussion

The lifting scheme is a new method of constructing biorthogonal wavelets, which makes the wavelet constructs free from the dependence of the Fourier transform. Before the chaotic encryption, the image data is wavelet transformed, and if one of the wavelet coefficients obtained after the transformation changes, the inverse of the wavelet transform is reflected in all the pixel points. This encryption effect is much better than using only chaotic sequence encryption. Single use of scrambling techniques or a single use of mixed-purchase sequences for encryption [16]. The traditional encryption algorithm does not change the pixel value of the image, only scrambles the position of the image pixel value, or only changes the pixel value. The correlation function of chaotic signals is similar to the correlation function of random signals and has the characteristics of impulse function [17–20]. In the process of algorithm simulation, the original image is scrambled in different areas, and different Baker mappings are used for each area. If other diffusion methods are used, the pixel values can be greatly changed, but the time complexity of the algorithm will be increased accordingly.

In order to make up for the shortcomings of the existing shrinkage function, a new curve shrinkage function is constructed. Make the estimates continuous at the threshold. With the increase of radial component, it can reach and exceed the true value. The shrinkage trend of adaptive non-linear curve is shown in Fig. 3.

It not only preserves the randomness and initial value sensitivity of Logistic, but also confuses the image with the image diffusion, which changes the pixel value of the image. From the definition of multi-resolution analysis, the standard orthogonal vector group of image encryption can be expressed as:

$$f(x) = sign[\omega^T x + b] \qquad (4)$$

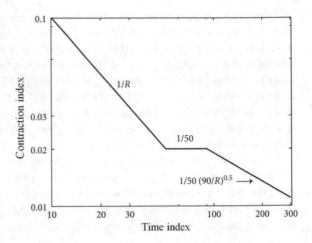

Fig. 3 Adaptive nonlinear curve shrinkage trend

There must be a two-scale equation:

$$w(t) = w_2 + (w_1 - w_2)\frac{T - t}{T} \tag{5}$$

Then the direct sum representation of the orthogonal complement closed subspace sequence is as follows:

$$e_j = -k \sum_{i=1}^{n} f_{ij} \ln f_{ij} \tag{6}$$

Take the formula:

$$W_j = 1 + k \sum_{i=1}^{n} f_{ij} \ln f_{ij} \bigg/ \sum_{j=1}^{m} \left(1 + k \sum_{i=1}^{n} f_{ij} \ln f_{ij}\right) \tag{7}$$

Then the wavelet corresponding to the scale function is:

$$W_j = d_j \bigg/ \sum_{j=1}^{m} d_j \tag{8}$$

In the method of changing the wavelet transform coefficient of a specific position by using a chaotic sequence, the image hiding technology is further used to realize the separate storage of the key and the encrypted image. Since the wavelet transform has a zooming effect and a local frequency locality, the detailed information of the image can be well preserved while removing noise, so that a high quality image can be reconstructed. The analytic hierarchy process is used to comprehensively evaluate the risk of digital image perception layer. Compare the relative importance of each factor in the same level with respect to the same factor of the previous level, and construct a pairwise comparison matrix. The statistical data results are shown in Table 3.

Table 3 Digital image perception layer risk comparison data results

	Residence time	Arrival rate	Views
Residence time	1	0.32	0.78
Arrival rate	0.87	1	0.59
Views	0.66	0.45	1

Chaotic phenomena are sensitive to the initial conditions. As long as the initial conditions are slightly different or slightly disturbed, the final state of the system will appear a huge difference binary sequence. When processing the wavelet signal, the wavelet can be transformed into continuous and discrete wavelet transform. It avoids the disadvantage that it is almost impossible to get the whole field data in the traditional manual analysis method of fringe pattern because of its huge amount of calculation. The high-speed computing ability of the computer is fully utilized, which greatly shortens the time needed for fringe analysis. After drawing the segmentation of the original signal, it is found that the wavelet transform has good time positioning performance. According to the simulation of the time and frequency signals of wavelet transform, it is found that the signal segments in the low frequency part are relatively blurred, while in the high frequency part, the original signal segments are very clear. There is no imaginary part in the result of wavelet continuous transform, because the wavelet transform uses a real function transform kernel. Wavelet transform is a reversible transform of the signal-preserving type, and the information of the original image or signal is completely retained in the coefficients of the wavelet transform.

The wavelet transform simply redistributes the energy of the original image or the letter one. Chaotic systems exhibit unpredictability, non-decomposability, but have certain regularity. In order to reflect the convergence speed and global optimal fitness value of the clustering and cluster head election algorithm for gateway selection optimization, the relationship between the reciprocal of the global optimal wavelet parameters and the number of iterations of the digital image is compared, as shown in Table 4.

In the practical application of image processing, because the image signal itself can be regarded as a two-dimensional signal, the image is decomposed into two-dimensional discrete wavelet transform. As the noise variance increases, the peak signal-to-noise ratio of each algorithm decreases, and the peak signal-to-noise ratio of the proposed algorithm decreases relatively slowly. And when the noise is constant, the signal-to-noise ratio of the improved algorithm is significantly higher than other

Table 4 Convergence rate under different inertia weights

Inertia weight	First round	Second round	Third round	Fourth round
0.8	3211	3353	3437	3477
0.16	4564	4434	4517	4529
This paper	5418	5575	4611	4739

algorithms. Figure 4 shows the mean square error of the dena noise of each algorithm under different noise intensities. Figure 5 is a comparison of the mean square
error of the IoT image after noise reduction for each algorithm under different noise
intensities.

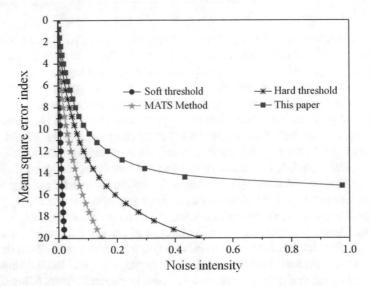

Fig. 4 Comparison of mean square error after denoising of each algorithm of lena image under
different noise intensities

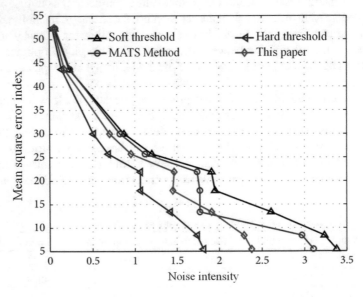

Fig. 5 Comparison of mean square error after noise reduction of IoT images under different noise
intensities

Classic shrink functions are hard threshold functions and soft threshold functions. Since the hard threshold function is discontinuous at the threshold, it is easy to cause edge oscillation of the reconstructed image. The soft threshold function cannot reach the true value of the wavelet coefficient, which easily causes the edge of the image to be blurred. For the input simulation parameters, the node arrival rate, the service rate, and the number of service stations per node are also included. The specific input parameters are shown in Table 5.

According to the scope division of the image encryption algorithm, it can be divided into airspace and frequency domain encryption. Among them, the encryption in the airspace has become the most used algorithm because of its fast speed and good encryption effect. The local variance is the principle that the noise contained in each high-frequency sub-band of each digital image decomposition layer is different after the digital image wavelet decomposition, and the noise variance is calculated on each high-frequency sub-band of each wavelet decomposition layer. The comparison of the average end-to-end delay between the gateway and the source node is shown in Fig. 6. At a rate at which different node locations change, the packet delivery rate between the gateway and the source node is compared as shown in Figure At different rate of change in node position (Fig. 7).

Table 5 Node parameter input

Arrival rate	Service rate	Number of desks
2	99	2
4	97	2
3	97	3
3	98	4

Fig. 6 Comparison of average end-to-end delay between gateway and source nodes

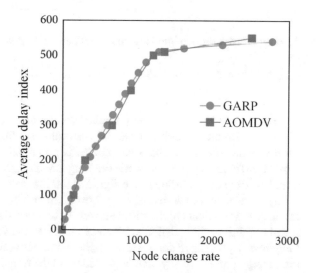

Fig. 7 Comparison of packet delivery rates between gateways and source nodes

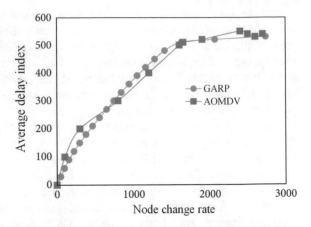

Any real orthogonal decomposition wavelet filter bank can realize image decomposition and synthesis, but not any decomposition can meet specific requirements. The fast algorithm for digital image two-dimensional discrete wavelet decomposition can be expressed as:

$$i_t^* = r^* + \pi_t + \alpha(\pi_t - \pi^*) + \beta(y_t - y_t^*) + \gamma e_t + \delta m_t) \tag{9}$$

When encrypting the system, the generation of the key stream has a great impact on the security of its encryption. The original image signal can be reconstructed by the low frequency approximation coefficient and three high frequency detail coefficients of wavelet decomposition. The reconstruction process can be expressed as:

$$i_t = (1 - \rho)i_t^* + \rho i_{t-1} + \xi_t \tag{10}$$

A noisy two-dimensional signal model can be expressed in the following form:

$$i_t = (1 - \rho)\left[r^* + \pi_t + \alpha(\pi_t - \pi^*) + \beta(y_t - y_t^*) + \gamma e_t + \delta m_t\right] + \rho i_{t-1} + \xi_t \tag{11}$$

Chaotic signals have a non-periodic, continuous broadband spectrum, noise-like characteristics, making it natural concealment. In practical applications, the short-wavelength of the short-wavelength is commonly used to extract the short-term high-frequency portion, while the long-wavelength is used to extract the long-term low-frequency portion of the signal. After the image is decomposed by integer lifting wavelet, some sub-band images are obtained. These sub-images have different frequency coefficients under different resolution conditions. The high sensitivity to initial conditions and small disturbances makes chaotic signals unpredictable for a long time. By sensitive dependence of chaotic system on initial values and parameters, a large number of keys can be provided. In the transformation results, the larger the scale, the lower the frequency, the smaller the scale, the higher the frequency. In the low frequency part, the frequency resolution is higher, but the time resolution is lower. In contrast, the frequency resolution is low, but it has a high time resolution. When the image edge is extended symmetrically, the reconstructed image edge is less real, which is conducive to obtaining high quality reconstructed image.

4 Conclusions

In this paper, an algorithm for generating chaotic encryption templates and scrambling sequences is established from two-dimensional Logistic maps. Secondly, the steps of image processing using two-level chaotic encryption are given. Finally, the method of recovering the encrypted image is given. The key sequence is generated by scrambling technology and chaotic system, and the coefficients in the wavelet transform domain are encrypted. Both confidential image information is protected without adding extra data. In the field of image processing, discrete wavelet transform is often used for image analysis and processing. After multi-level wavelet decomposition of the image, the low-frequency components of the wavelet decomposition are extracted for diffusion and reconstruction operations. Although the lifting wavelet can perform lossless decryption on image level decomposition, the image after decryption still has pixel point loss when performing advanced decomposition. With the continuous advancement of image encryption technology, more and more practical image encryption technologies will emerge.

References

1. Bhatnagar G, Wu QMJ, Raman B (2013) Discrete fractional wavelet transform and its application to multiple encryption. Inf Sci 223:297–316
2. Zheng P, Huang J (2013) Discrete wavelet transform and data expansion reduction in homomorphic encrypted domain. IEEE Trans Image Process 22(6):2455–2468
3. Tedmori S, Alnajdawi N (2014) Image cryptographic algorithm based on the Haar wavelet transform. Inf Sci 269(11):21–34
4. Asatryan D, Khalili M (2013) Colour spaces effects on improved discrete wavelet transform-based digital image watermarking using Arnold transform map. IET Signal Process 7(3):177–187
5. Hu Y, Xie X, Liu X et al (2017) Quantum multi-image encryption based on iteration arnold transform with parameters and image correlation decomposition. Int J Theor Phys 56(7):2192–2205
6. Mehra I, Fatima A, Nishchal NK (2018) Gyrator wavelet transform. IET Image Process 12(3):432–437
7. Chun-Lai L, Hong-Min L, Fu-Dong L et al (2018) Multiple-image encryption by using robust chaotic map in wavelet transform domain. Optik 171:277–286
8. Tong XJ, Wang Z, Zhang M et al (2013) A new algorithm of the combination of image compression and encryption technology based on cross chaotic map. Nonlinear Dyn 72(1–2):229–241
9. Hamdi M, Rhouma R, Belghith S (2016) A selective compression-encryption of images based on SPIHT coding and chirikov standard MAP. Signal Process 131:514–526

10. Multilevel Encrypted Text Watermarking on Medical Images Using Spread-Spectrum in DWT Domain. Wirel Pers Commun **83**(3), 2133–2150 (2015)
11. Gong LH, He XT, Cheng S et al (2016) Quantum image encryption algorithm based on quantum image XOR operations. Int J Theor Phys 55(7):3234–3250
12. Beltrán del Río L, Gómez A, José-Yacamán M (1991) Image processing in TEM using the wavelet transform. Ultramicroscopy 38(3–4):319–324
13. Rajakumar K, Arivoli T (2016) Lossy image compression using multiwavelet transform for wireless transmission. Wireless Pers Commun 87(2):315–333
14. Lu H, Li Y, Uemura T, Kim H, Serikawa S (2018) Low illumination underwater light field images reconstruction using deep convolutional neural networks. Futur Gener Comput Syst 82:142–148
15. Wang, M., Zhou, S., Yang, Z., et al.: Image fusion based on wavelet transform and gray-level features. J. Mod. Opt. 1–10 (2018)
16. Sophia PE, Anitha J (2017) Contextual medical image compression using normalized wavelet-transform coefficients and prediction. IETE J Res 63(5):1–13
17. Serikawa S, Lu H (2014) Underwater image dehazing using joint trilateral filter. Comput Electr Eng 40(1):41–50
18. Lu H, Li Y, Mu S, Wang D, Kim H, Serikawa S (2018) Motor anomaly detection for unmanned aerial vehicles using reinforcement learning. IEEE Internet Things J 5(4):2315–2322
19. Lu H, Li Y, Chen M, Kim H, Serikawa S (2018) Brain Intelligence: go beyond artificial intelligence. Mob Netw Appl 23:368–375
20. Lu H, Wang D, Li Y, Li J, Li X, Kim H, Serikawa S, Humar I (2019) CONet: A Cognitive Ocean Network. In Press, IEEE Wireless Communications

High Level Video Event Modeling, Recognition and Reasoning via Petri Net

Zhijiao Xiao, Jianmin Jiang, and Zhong Ming

Abstract A Petri net based framework is proposed for automatic high level video event description, recognition and reasoning purposes. In comparison with the existing approaches reported in the literature, our work is characterized with a number of novel features: (i) the high level video event modeling and recognition based on Petri net are fully automatic, which are not only capable of covering single video events but also multiple ones without limit; (ii) more variations of event paths can be found and modeled using the proposed algorithms; (iii) the recognition results are more accurate based on automatic built high level event models. Experimental results show that the proposed method outperforms the existing benchmark in terms of recognition precision and recall. Additional advantages can be achieved such that hidden variations of events hardly identified by humans can also be recognized.

Keywords Automated video event modeling · Video event recognition · Video event reasoning · Petri net

1 Introduction

With the rapid development of artificial intelligence [1, 2], computerized video content analysis is moving towards high level semantics based approaches, where video event recognition and reasoning remains to be one of the actively researched topics over the past decades. To narrow the gap between low level visual features and high level semantics, existed methods focus on two levels of video event analysis. The

Z. Xiao (✉) · J. Jiang · Z. Ming
College of Computer Science & Software Engineering, Shenzhen University, Shenzhen 518060, China
e-mail: cindyxzj@szu.edu.cn

J. Jiang
e-mail: jianmin.jiang@szu.edu.cn

Z. Ming
e-mail: mingz@szu.edu.cn

© Springer Nature Switzerland AG 2021
H. Lu (ed.), *Artificial Intelligence and Robotics*,
Studies in Computational Intelligence 917,
https://doi.org/10.1007/978-3-030-56178-9_6

low level is to recognize atomic actions. Researches on this area are often key-frame based. Global and local features are extracted from those key frames and semantic concept classifiers are applied to capture crucial patterns for event recognition. There are many ways to extract low level features which have been successfully applied in many areas [3–5]. Hasan et al. [6] propose a framework for continuous activity learning using deep hybrid feature models and active learning. Samanta et al. [7] use three-dimensional facet model to detect and describe space time interest points in videos. Those methods can extract low level semantics for action recognition and event analysis based on key actions or scenes. They give no or less consideration about temporal and logical relations among actions when they are used to classify events. Some improvement researches are done. Wang et al. [8] propose a new motion feature to compute the relative motion between visual words and present approaches to select informative features. Cui et al. [9] propose a novel unsupervised approach for mining categories from action video sequences. They use pixel prototypes quantized by spatially distributed dynamic pixels to represent video data structuration. Abbasnejad et al. [10] present a model based on the combination of semantic and temporal features extracted from video frames, that is able to detect the events with unknown starting and ending locations.

The work in this paper focuses on the high level of video event recognition. The high level is conducted on the results of the low level to recognize events with complex action sequences. Veeraraghavan et al. [11] present semi supervised event learning algorithms. The models of events are represented as stochastic context-free grammars. Kitani et al. [12] create a hierarchical Bayesian network by combining stochastic context-free grammar and Bayesian network. They apply the network on action sequences via deleted interpolation to recognize events. Shet et al. [13] use Prolog based reasoning engine to recognize events from log of primitive actions and predefined rules. Song et al. [14] present a multi-modal Markov logic (ML) framework for recognizing complex events. Liu et al. [15] present an interval-based Bayesian generative network approach to model complex activities by constructing probabilistic interval-based networks with temporal dependencies in complex activity recognition. Song et al. [16] present a framework for high-level activity analysis which consists of multi-temporal analysis, multi-temporal perception layers, and late fusion. The method can handle temporal diversity of high-level activities. Nawaz et al. [17] propose a framework for predictive and proactive complex event reasoning which processes, integrates, and provides reasoning over complex events using the logical and probabilistic reasoning approaches. Skarlatidis et al. [18] present a system for recognizing human activity given a symbolic representation of video content. They use a dialect of the Event Calculus for probabilistic reasoning. Cavaliere et al. [19] employ semantic web technologies to encode video tracking and classification data into ontological statements which allow the generation of a high-level description of the scenario through activity detection. By semantic reasoning, the system is able to connect the simple activities into more complex activities. Jorge et al. [20] propose a predictive method based on a simple representation of trajectories of a person in the scene which allows a high level understanding of the global human behavior. Their method does not need predefined models and rules to evaluate behaviors.

Since Castel et al. [21] introduce Petri nets for high level representation of image sequences, Petri nets and its variations are widely used in modeling video events for their detection and recognition. Petri net is a powerful event model tool that supports the representation of high level events. While places denote different states of objects inside videos, transitions represent switches of states that are usually caused by primitive actions performed by the tracked objects. When such a model is used to recognize an event, each tracked object will be modeled as a token moving in the Petri net model according to its action sequence. If any token reaches the end place of the event model, the event is claimed to have happened. During the tracking process, event reasoning can be done to predict which event has the biggest possibility to happen.

Albanese et al. [22] propose an extended Probabilistic Petri Nets. They present the PPN-MPS algorithm to find the minimal sub-videos that contain a given activity with a probability above a certain threshold. Ghanem et al. [23, 24] address the advantages of using Petri nets for event recognition. They propose a framework which provides a graphical user interface for user to define objects and primitive events, and then expresses composite events using logical, temporal and spatial relations. Lavee et al. [25] propose the Particle Filter Petri Net to model and recognize activities in videos. They also propose a method to transform semantic descriptions of events in formal ontology languages to Petri net event models [26]. The surveillance event recognition framework they proposed uses a single Petri net for recognition of event occurrences in video, which allows modeling of events having variances in duration and predicting future events probabilistically [27]. Borzin et al. [28] present video event interpretation approach using GSPN. Through adding marking analysis into a GSPN model, their methods provide better scene understanding and next marking state prediction using historic data. Najla et al. [29] present an approach to automatically detect abnormal high-level events in a parking lot. A Petri net model is used to describe and recognize high-level events or scenarios that incorporate simple events with temporal and spatial relations. Hamidun et al. [30] translate the event sequence in the crossing scenario to the PN model. The combined effects of spatial and temporal information were analyzed using the steady state analysis built in the model. They point out that modeling with Petri Nets also allows the development of model in hierarchical structure. Szwed [31, 32] proposes Fuzzy Semantic Petri Nets (FSPN) as a tool aimed at solving video event modeling and recognition problems. Linear Temporal Logic is used as a language for events specification and FSPN is used as a tool for recognition. SanMiguel et al. [33] use Petri nets in the long-term layer, which is in charge of detecting events with a temporal relation among their counterparts. They extend the basic PN structure to manage uncertainty obtained by the sub-events.

Existing researches focus on event recognitions, video event modeling is often ignored and remains as one of the unsolved research problems. Existing efforts are primarily limited to manual modeling approaches, including knowledge-based or rule-based schemes through semantics extractions. Although significant progress has been achieved, it is stated by many researchers [34–37] that automatic event modeling is still a challenge. In this paper, we introduce a high level video event modeling, recognition and reasoning approach based on Petri net to forward the existing state of the arts on Petri net based video event recognition, providing a pioneering framework for computerized high level video content interpretation, analysis and understanding. To this end, our main contribution can be highlighted as:

(i)   We systematically propose a Petri net based high level video event description model, which can be expanded for describing any high level video event for video content analysis and semantics organization;

(ii) We present algorithms which can directly build up a video event model from the labeled video training dataset automatically without any intermediate entities such as ontologies etc.

(iii) More variations of event paths can be captured and the recognition results can be more accurate based on automatic built high level event models.

(iv) The rest of the paper is organized as follows. Section 2 presents some concepts of the Petri net based video event modeling, in order to pave the way for our proposed work. Section 3 describes our proposed algorithms for high level video event modeling and recognition based on Petri net, and Sect. 4 reports the experimental results. Comparative analysis of the results is also included in this section to evaluate the performance of the proposed methods, and finally, a conclusion and proposals for future research are addressed in Sect. 5.

2   Petri Net Based Video Event Modeling

Petri net is a directed graph constructed with four essential elements: place, transition, arc and token. While the first three elements are used to model the static structures, token is designed to reflect the dynamic states of a Petri net. For a detailed Petri net introduction, we refer to [38].

Let the places representing possible states of tracked objects, the transitions representing possible primitive actions of tracked objects, and tokens standing for tracked objects, we can define a single event model (SE-Tree) as follows based on the concept of a classical Petri net.

Definition 1 (*SE-Tree*) A *PN* is a SE-Tree if and only if:

(1) There exists one and only one source place $p_0 \in P$, and $\cdot p_0 = \Phi$, and for $\forall$ $p \in P\text{-}\{p_0\}$, $|p| = 1$;

(2) If p is a leaf place, $p \cdot = \Phi$, otherwise $|p| \geq 1$;

(3) $P_e \subset P$ is a finite set of end places. An end place denotes a final state of the object that conducted an event.

(4) For $\forall\, t \in T$, $|t| = 1$ and $|t| = 1$.

Since there exist many uncertainties inside an event, a SE-Tree often needs to model all its possible variations, and each of such variations is referred to as an instance. To provide efficient and effective coverage of all the possible uncertainties, we introduce the concept of a path to describe the route of an event instance.

Definition 2 (*Path*) A Path, $path = <p_0, t_0, p_1, \ldots, t_i, p_{i+1} \ldots, t_{m-1}, p_m>$, is a sequence of nodes which connect the source place p_0 to one of the end place $p_m (p_m \in P_e)$.

A SE-Tree modelling all paths of an event has a tree structure. We choose the tree structure other than net because there is less ambiguity and inaccuracy. Based on the concept of SE-Tree as described above, a Petri net based multi event model is defined as follows.

Definition 3 (*ME-Tree*) A *PN* is a ME-Tree if and only if:

(1) Each $p \in P$ has an attribute called *par_event*. The value of this attribute is the ids of all possible events that the place participated;
(2) Each $p \in P_e$ has an attribute called *end_event*. The value of this attribute is the id of the most possible event that the place is the end place.

3   High Level Video Event Model Building

The process of high level video event model building is depicted in Fig. 1.

To build a high level event model, a certain amount of video segments containing specific high level events should be prepared as the training dataset. Here we suppose that all objects and their primitive actions and states are recognized and target events are labeled. Our work focuses on the last two steps. The symbols used are explained in Table 1.

3.1   Single Event Model Building

The places and transitions of SET_k are created based on following rules.

Rule 1: A process of creation will be fired for f_u in a video segment if and only if:

(1) $\exists v (FGF(f_u, o_v, event) = e_k)$, and
(2) $FGO(o_v, state) \neq FGF(f_u, o_v, state)$, and

(3)
$$\neg \exists j (p_i = FGO(o_v, place) \wedge t_{ij} \in p_i \bullet \wedge p_j \in t_{ij} \bullet \wedge FGP(p_j, label)$$
$$= FGF(f_u, o_v, state))$$

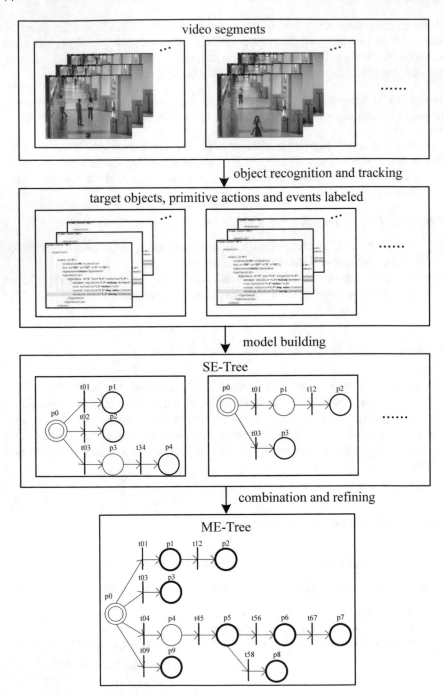

Fig. 1 Process of video event model building

Table 1 List of symbols and their descriptions

Symbols	Descriptions
P	The set of places which includes all objects' available states
p_i	The ith place, $i = 0,\dots, n-1$
P_e	$P_e \subset P$ is a finite set of end places
T	The set of transitions which includes primitive actions or changes of objects' states
t_{ij}	Transition connect p_i–p_j
$\bullet\, p_i$	$\{t_{ji}\| < t_{ji}, p_l > \in T \times P\}$ is the set of input transitions of p_i
$p_i\bullet$	$\{t_{ij}\| < p_i, t_{ij} > \in P \times T\}$ is the set of output transitions of p_i
$\bullet\, t_{ij}$	$\{p_i\| < p_i, t_{ij} > \in P \times T\}$ is the input place of t_{ij}
$t_{ij}\bullet$	$\{p_j\| < t_{ij}, p_j > \in T \times P\}$ is the output place of t_{ij}
o_v	The vth token, which is a tracked object
e_k	The kth event, $k = 0, \dots, m-1$
f_u	The uth frame of a video segment
SET_k	The SE-Tree for e_k
TP_i	Number of tokens that pass by p_i
TP_{ik}	Number of tokens that pass by p_i and conduct event e_k
TEP_i	Number of tokens that end in p_i
TEP_{ik}	Number of tokens that end in p_i and conduct event e_k
TT_{ij}	Number of tokens that pass by t_{ij}
TT_{ijk}	Number of tokens that pass by t_{ij} and conduct event e_k
PE_i	Probability of an token ended in p_i
PP_{ik}	Probability of a token reached in p_i who will conduct event e_k
PEE_{ik}	Probability of a token ended in p_i who may conduct event e_k
PT_{ij}	Fire probability of t_{ij} who connects p_i–p_j for a token that passes by p_i and not ends in p_i
S_{vu}	State of o_v in f_u
SC_v	Current state of o_v
$FGF\,(f_u, o_v, w)$	The function which can get the value of the attribute w of o_v in f_u
$FGP\,(p_i, w)$	The function which can get the value of the attribute w of p_i
$FGO\,(o_v, w)$	The function which can get the latest value of the attribute w of o_v
$FSO\,(o_v, w, a)$	The function which can set the value of the attribute w of o_v–a
$FSP\,(p_i, w, a)$	The function which can set the value of the attribute w of p_i–a

Condition (1) requires a tracked object o_v is conducting e_k; (2) means that there is a change of o_v's state in f_u; (3) denotes that there is no output transitions t_{ij} of p_i(the place that o_v stay currently), whose output place p_j has the same label as the new state of o_v.

If a tracked object o_v is conducting event e_k, all its states and switches of states will be modeled. The creation will first trigger the creation of a new place p_j which describes the new state of the object inside f_u. After the new place is created, a new transition t_{ij} will be created to connect the two places p_i and p_j, while p_i denotes the old state and p_j denotes the new state of the object o_v. After the creation, the token representing o_v will be moved from p_i to p_j, and the values of o_v's attributes of o_v will be updated.

Rule 2: A place is marked as an end place in SET_k if and only if the place is the final state of an object conducted e_k.

In order to support the certainty reasoning of video event, the probability distribution over the places in an event model needs to be learned to estimate the likelihood an event's occurrence. The number of tokens that once stay in each place and the number of tokens that fire each transition will be counted. If a place is marked as an end place, it will count the number of tokens who end in this place and which event each token ends. Specific values of those numbers are learned for generating the probability of the event's occurrence.

Out of the above rules and formulas, the algorithm for SE-Tree Building can be described as follows.

Algorithm 1: SE-Tree Building (Single Even Modeling)

Input: e_k: Event needed to be modeled

FQ: A queue of frames of a video segment containing the event e_k

Output: SET_k: A SE-Tree of Event e_k

Initialize empty P, T, $P \times T$ and $T \times P$ of se

Create a place p_0 as the source place and add it to P

$FSP(p_0, label,$ "unkown")   //initialize the label of p_0

while FQ not empty do

  $f_u = FQ.pop()$

    for each object o_v conducting event e_k in f do

      if o_v just appear

        create a new token o_v

        add o_v to p_0

        $FSO(o_v, place, p_0)$ //initialize the stay place of o_v

        $FSO(o_v, state,$ "unkown")   //initialize the state of o_v

        $TP_{0k}++$

      end

      $p_i = FGO(o_v, place)$

      if rule 1 is satisfied

        create a new place p_j and $FSP(p_j, label, FG_f(f_u, o_v, state))$ //set the label of p_j

        add p_j to P

        create a new transition t_{ij}, add t to T

        create a new arc pt from p_i to t_{ij}, add pt to $P \times T$

        create a new arc tp from t_{ij} to p_j, add tp to $T \times P$

      else if only the third condition in rule 1 is broken

        find p_j that breaks the third condition in rule1

      end

      $FSO(o_v, place, p_j)$ //update the stay place of o_v

      $FSO(o_v, state, FG_p(p_j, label))$ //update the state of o_v

      TP_{jk}++

      TT_{ijk}++

      if o_v ends in p_j

        TEP_j++

        TEP_{jk}++

      end

      if rule 2 is satisfied

        add p_j to P_e

      end

    end

  end

The computational complexity of the algorithm for SE-Tree building depends on the number of frames in a video segment and the number of tracked objects conducting event e_k in each frame. Given n_i represents the number of objects conducting event e_k in the ith frame of a video($i = 1, 2, \ldots m$, m is the number of frames in a video segment), the computational complexity of Algorithm 1 is $O\left(\sum_{i=1}^{m} n_i\right)$. Give N clips of videos, the computational complexity of SE-Tree building would be $O\left(\sum_{j=1}^{N} \left(\sum_{i=1}^{m_j} n_i\right)\right)$, where m_j is the number of frames in the jth($j = 1, 2, \ldots N$) video segment.

3.2 Multi Event Model Building

A ME-Tree can be built up by combining and refining several given single event models. Not all places in all SE-Trees are added to ME-Tree. Those duplicated places will be composed into one place to simplify the combined model. Let p' be the current place considered in SET_k, p_i is the corresponding place of p' in MET, a new place will only be created in MET according to the following rule.

Rule 3: A process of creation will be fired for *MET* if and only if:

(1) $\exists t'\left(t' \in T_k \land t' \in p'\cdot\right)$, and

(2) $\neg\exists j\left(t_{ij} \in T \land p_j \in P \land \left(t_{ij} \in p_i \cdot \land t_{ij} \in \cdot p_j\right)\right.$
$\left.\land\left(FGP\left(p_j, label\right) = FGP\left(t'\cdot, label\right)\right)\right)$

Condition (1) considers if there is a transition which is an output transition of current place *p'*. If condition (1) is satisfied, for an output place of *t'*, condition (2) checks if the corresponding place has existed in *MET* that is derived from the same source place p_i (*p'* in SET_k is modeled as p_i in *MET*). According to rule 3, those transitions connecting the same source and destination places will only be added to the multi event model once.

The algorithm for ME-Tree building is described as follows.

Algorithm 2: ME-Tree Building (Multi Events Modeling)

Input: *SET*: A set of single event models, each of which models a single event.

Output: *MET*: A ME-Tree

Initialize empty P, T, $P \times T$ and $T \times P$ of *MET*

Create a place p_0 as the source place and add it to P

$p = p_0$

for each $SET_k \in SET$ do

   $p' = p'_0$ // $p'_0 \in SET_k$

   if $p'\cdot \neq \Phi$

      creation(p_0, p')

   end

end

---

creation(p_i, p')

for each $t' \in p'\cdot$

   if rule 3 is satisfied,

      create a new place p_j

      add p_j to P

add the information of *se* to update the *par_event* of p_j

create a new transition t_{ij}, add t_{ij} to T

create a new arc *pt* from p_i to t_{ij}, and add *pt* to $P \times T$

create a new arc *tp* from t_{ij} to p_j, and add *tp* to $T \times P$

else

find p_j that breaks the second condition in rule 3

end

$p' = t' \cdot$

if $p' \in P_e$'

add p_j to P_e

add the information of *se* to update the *end_event* of p

search other models in *SE*

if there is no path through which a token can reach p'

break;

end

end

if $p' \cdot \neq \Phi$

creation(p_j, p')

end

end

According to Algorithm 2, the ME-Tree will be simplified by eliminating those unnecessary nodes as soon as the checking list of *end_event* is narrowed to a unique one.

The computational complexity of the algorithm for ME-Tree building depends on the number of SE-Trees and the number of places and transitions in each SE-Tree. Given n_i represents the number of transitions in the *i*th SE-Tree($i = 1, 2, ...m, m$ is the number of SE-Tree), the computational complexity of Algorithm 2 is $O(\sum_{i=1}^{m} n_i)$.

Based on the numbers learned during single event modeling, the probabilities for event reasoning can be calculated as follows.

$$PP_{ik} = \begin{cases} PEE_{ik}, p_i \cdot = \emptyset \\ PEE_{ik} \times PE_i + (1 - PE_i) \times \sum_{t_{ij} \in p_i \cdot} \left(PT_{ij} \times PP_{jk} \right), others \end{cases} \quad (1)$$

where

$$PT_{ij} = \frac{TT_{ij}}{TP_i - TEP_i} \quad (2)$$

$$PE_i = \frac{TEP_i}{TP_i} \quad (3)$$

$$PEE_{ik} = \frac{TEP_{ik}}{TEP_i} \quad (4)$$

$$TT_{ij} = \sum_{k=0}^{m-1} TT_{ijk} \tag{5}$$

$$TP_i = \sum_{k=0}^{m-1} TP_{ik} \tag{6}$$

$$TEP_i = \sum_{k=0}^{m-1} TEP_{ik} \tag{7}$$

$$\sum_{j=0}^{n-1} PT_{ij} = 1 \tag{8}$$

$$\sum_{k=0}^{m-1} PEE_{ik} = 1 \tag{9}$$

$$\sum_{k=0}^{m-1} PP_{ik} = 1 \tag{10}$$

This information about event reasoning is added as attributes' values of places and transitions of a ME-Tree, which can be used to make predictions.

3.3   Video Event Recognition and Reasoning

A token in a ME-Tree denotes a tracked object. The events that tracked objects in a video segment have conducted or will conduct can be recognized by observing the distribution of tokens in the ME-Tree. A marking is a distribution of tokens over the set of places in a ME-Tree. It changes along with the firing of transitions. A transition is enabled if and only if there are tokens in the input place of the transition as defined in rule 4.

Rule 4: t_{ij} is enabled if and only if $M(p_i) \geq 0 \wedge G(t_{ij}) = true$.

$M(p_i)$ is the number of tokens stay in p_i under marking M. $G(t_{ij})$ denotes the set of all the guard functions on transition t_{ij} for firing. Here, the main guard function for firing a transition is a state change caused by a primitive action of an object or a group of objects.

A change of the vth object's state is detected in current frame, the transition will be fired. The token representing the vth object will be moved from p_i to p_j with the firing of t_{ij}. After the firing of transition t_{ij}, a new marking is generated according to rule 5.

Rule 5: t_{ij} is fired for the vth object in the uh frame and the marking M is replaced by a new marking M' produced according to Eq. (11) and (12).

$$M'(p_i) = M(p_i) - 1 \qquad (11)$$

$$M'(p_j) = M(p_j) + 1 \qquad (12)$$

An instance of an event is happened if there is a token (an object) reached one of the end places.

The probabilities for event reasoning can be updated and deduced for each object in each frame. The algorithm for event recognition based on a ME-Tree can be described as follows.

Algorithm 3: MER(Multi Events Recognition)
***

Input: ES: A set of events needed to be detected

MET: A ME-Tree

FQ: A queue of frames of a video segment with recognized objects and state of each object in each frame is recognized

Output: EN: An array contains numbers of happened events

for each event $e_k \in ES$ do

    $EN_k = 0$

end

while FQ not empty do

    $f_u = FQ.pop()$

    for each object o_v in f_u do

        if o_v just appear

        create a new token $T(o_v)$

        add $T(o_v)$ to p_0 and initialize the attribute values of o_v

        end

    end

```
for each t_{ij} ∈ T do
    if t_{ij} is enabled according to rule 4
    fire t_{ij} and change marking according to rule 5
    update attribute values of objects enabled the firing
        for each object o_v enabled the firing in f_u do
            FSO(o_v, pos_event, FGP(FGO(o_v, place), par_event))
            //set the possible event of o_v
            if FGO(o_v, place) ∈ P_e  //o_v reached an end place
                FSO(o_v, event, FGP(FGO(o_v, place), end_event))
                //set the happened event of o_v
            end
        end
    end
end
end
for each object o_v do
    if FGO(o_v, event) is empty
        FSO(o_v, event, FGO(o_v, MOST(pos_event)))
        // set the happened event of o_v as the most possible event
    end
    for each event e_k == FGO(o_v, event) do
        EN_k++
    end
end
return EN
```

The computational complexity of the algorithm mainly depends on the number of frames in a video segment, the number of objects in each frame. Given n_i represents the number of objects in the ith frame of a video ($i = 1, 2,...m$, m is the number of frames in a video segment), the computational complexity of Algorithm 3 is $O\left(\sum_{i=1}^{m} n_i\right)$.

4   Experimental Results and Discussion

In this section, we report experimental results to demonstrate the performance of the proposed methods. We use CAVIAR [39] and VIRAT [40] datasets in consideration that the two datasets contain video segments of high level events other than atomic actions or primitive activities which are not discussed here.

4.1 Experiments on CAVIAR Dataset

We select the published work [27] using manual event model and the published work [20] using non-Petri net model to compare, both of them use the CAVIAR dataset as the experimental dataset. The CAVIAR dataset contains 52 clips of videos of a shopping center in Lisbon. This set of sequences contains 1500 frames on average. The ground truth and labels are provided. Half of videos in a dataset are randomly chosen as the training dataset, and the other half are used as the test dataset.

4.1.1 Results of Automatic Single Event Model Building

Seven different contexts are considered here which are numbered and referred to as given in Table 2. The ground truth information, including context and situation information provided by the CAVIAR dataset, is used to build SE-Tree for each high level event, from which the context of each object is treated as a video event, and the situation information is used to define the contexts. Seven SE-Trees built for those events are shown in Fig. 2.

4.1.2 Results of Automatic Multi Events Model Building

After all the SE-Trees are built from the training dataset, algorithm for building ME-Tree will be called to create a ME-Tree for all events involved. The ME-Tree built up in this way is depicted in Fig. 3. The source place and the end place are marked in Fig. 3. The values of attribute PT_{ij} of transitions are also labeled in Fig. 3. Other details are omitted here.

Compared to the model depicted in [27], our model captures more hidden variations of events. For example, there is no direct unknown to shop enter sequence for "shop enter" in the model presented in [27], which is captured by our method. As a result, precision rates and recall rates for event recognition are both improved. The reason is that there is less manual intervention during the building process of event models using the proposed method, which can find some hidden variations of events that are often ignored by human.

As stated in Sect. 3.2, the complexity of ME-tree depends on two factors: the number of SE-Trees and the number of places and transitions in each SE-tree. A

Table 2 Seven contexts

Id	Event	Id	Event
0	Browsing	4	Windowshop
1	Immobile	5	Shop exit
2	Walking	6	Shop reenter
3	Shop enter		

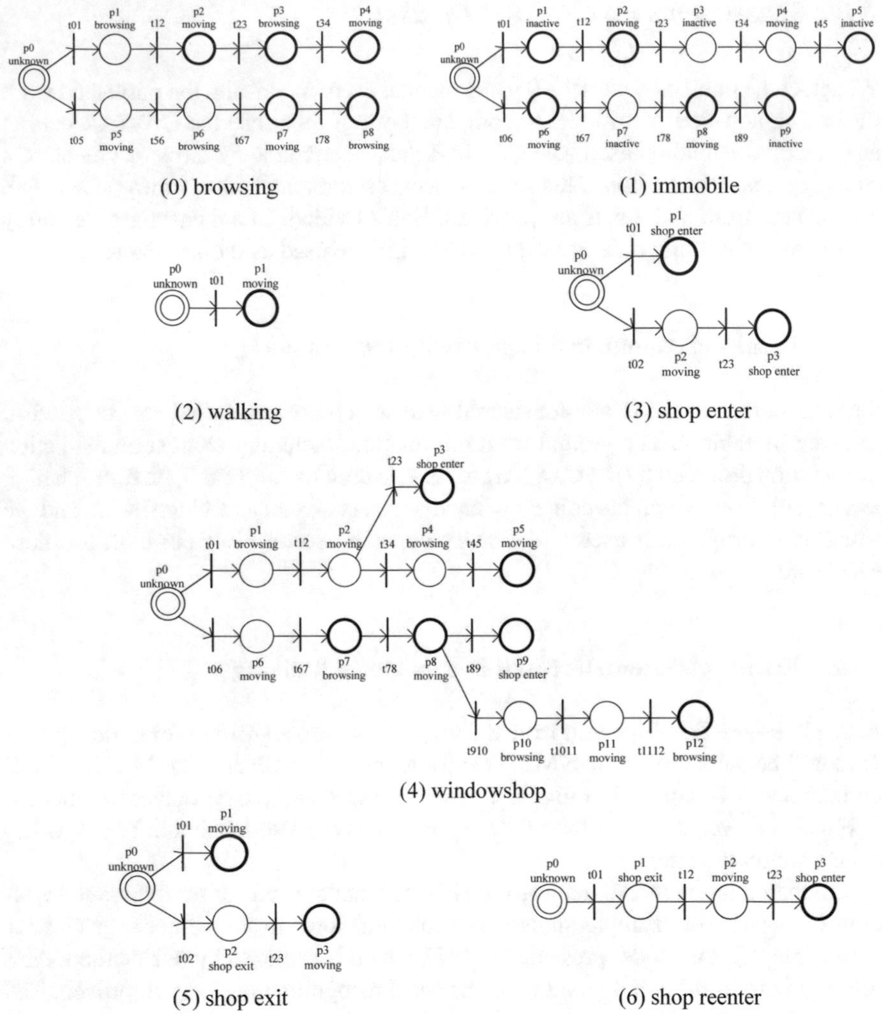

Fig. 2 SETs for seven events (◎ is the source place, ○ is the end place)

ME-Tree will combine and refine SE-Trees. If SE-Trees have similar or same paths or sub-paths, the ME-Tree will combine them, which decreases the scale of the model. But under this circumstance, event recognition will be more difficult due to the often conflicts.

If SE-Trees contain many different paths from each other, the complexity of ME-tree will increase since there are few nodes can be combined. But event recognition will be much easier. If those paths have unique actions to distinguish themselves from others, then the rest nodes of those paths after the unique actions can be eliminated according to Algorithm 2, which can bring down the complexity of the model.

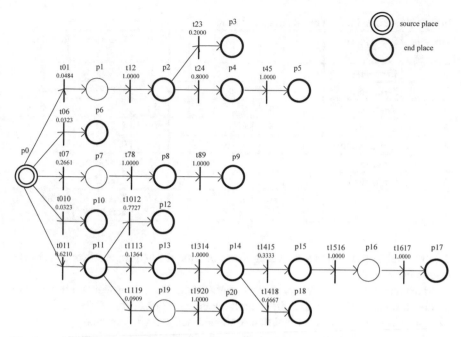

Fig. 3 ME-tree for CAVIAR contexts

However, if the unique actions appear at the end or near the end of the path, then there are not too many nodes left to be eliminated.

Another limitation of the proposed framework is that it only considers events that involve one object. But there are many events in reality that involves more than one object. To model this kind of event, extensions should be done on the proposed framework.

4.1.3 Results of Video Event Recognition Based on Different Models

A token will be created and added to the ME-Tree for each object. According to the firing rules described in Sect. 3.3, token will be moved from place to place. An instance of an event is happened if there is a token reaches one of the end places. Table 3 shows the recognition results of our proposed methods applied to the test dataset.

As seen in Table 3, there exist some false negatives and false positives, especially for "browsing" and "windowshop". The main reason is that they are similar events, which lead to similar sub-paths of the two events in the model. Actually, it is also very hard for human to distinguish the two events. When a conflict occurs, we choose the event with the largest probability according to our model building algorithm.

In order to compare to the results reported in [27] on the same test dataset, we conduct event recognition on the whole dataset(including training dataset and test

Table 3 Event classification based on automatic model

Event id	Event name	0	1	2	3	4	5	6	Total	Precision rate	Recall rate
0	**Browsing**	1				2			3	0.3333	0.3333
1	**Immobile**		6						6	1.0000	1.0000
2	**Walking**			48					48	0.9412	1.0000
3	**Shop enter**			3	15				18	1.0000	0.8333
4	**Windowshop**	2				5			7	0.7143	0.7143
5	**Shop exit**						25		25	1.0000	1.0000
6	**Shop reenter**							4	4	1.0000	1.0000
Total		3	6	51	15	7	25	4	111	**Overall accuracy**	0.9369

dataset). Table 4 compares the recognition results of our proposed method to the results listed in [27].

Since our model captures more event paths, there is an improvement in terms of precision rate, recall rate, and the overall accuracy. The results of "browsing" and "windowshop" are also the worst due to the similar paths in the model. According to the experimental results, we can draw a conclusion that the more unique actions the events have, the more accuracy the results are. The reason is less similar paths will exist among those events in the model if each event has its own unique actions.

We also conduct a series of experiments to compare the sensitivity, specificity, and accuracy of Petri net based methods and non-Petri net based method [20]. The definitions of the three performance indicator can be found in [20].

As shown in Table 5, our method outperforms the other two methods, while both Petri net based methods outperform the non-Petri net based one. The main reason is the method in [20] uses a pattern recognition approach assigning an event for just a

Table 4 Comparison of recognition results

Event id	Precision rate		Recall rate	
	Result reported in [27]	Result of our method	Result reported in [27]	Result of our method
0	0.0000	0.5714	0.0000	0.5714
1	1.0000	1.0000	0.1250	0.8750
2	0.8333	0.9167	0.9740	1.0000
3	0.9630	1.0000	0.9286	0.9464
4	0.5455	0.8000	0.8000	0.8000
5	0.9500	1.0000	0.9661	0.9661
6	0.8333	1.0000	1.0000	1.0000
Overall accuracy		**Result reported in [27]**		**0.8638**
		Result of our method		**0.9447**

Table 5 Comparison of recognition performance

Performance	Non-Petri net based method	Petri net based methods	
	Result reported in [20]	Result reported in [27]	Result of our method
Sensitivity	0.772	0.864	0.945
Specificity	0.962	0.978	0.991
Accuracy	0.935	0.962	0.984

trajectory instead of a predefined, semantic model of event. Hence, the semantic gap between the input and the output could be very large.

4.2 Experiments on VIRAT Dataset

VIRAT Video Dataset release 1.0 is a collection of video clips recording people doing different actions in a parking land. The ground truth information including event and situation information are not provided. So we labeled the video segments in the VIRAT dataset manually. 57 clips of videos from training dataset are used here as training data. 102 clips of videos from test dataset are used as test data.

We use the videos in the training dataset to build model for each event. After the ME-Tree including all paths of all events has been build, the videos in the test dataset are used to conduct video recognition. As shown in Table 6, the precision rate and the recall rate are quite acceptable.

5 Conclusions

A framework for high level video event modeling, recognition and reasoning is proposed in this paper to facilitate the video content analysis. Experimental results show that the proposed method can capture variants of paths inside a high level event and achieve improved performances in comparison with the benchmark. The accuracy of event recognition is improved based on the proposed automatic model in comparison with that based on manual models.

With a simple extension of our method, it can be applied to build models of multiple high level events that involve more than one object. In addition, the proposed method provides an efficient tool for managing video content analysis and interpretation in terms of high level events, leading to potential applications for high level understanding of the video content by computers.

Table 6 Event classification based on automatic model

Event name	1	2	3	4	5	6	7	8	9	10	Total	Precision rate	Recall rate
Chatting	7						1				8	0.4118	0.8750
Riding		45									45	1.0000	1.0000
Running			6		2						8	1.0000	0.7500
Skating				2							2	1.0000	1.0000
Walking	2				486		6				494	0.9818	0.9838
Delivering	3			1		17	11	5	2		39	1.0000	0.4359
Driving	1			1			18				20	0.4737	0.9000
Parking							2	17	3		22	0.5313	0.7727
Taking stuff	1				5		3	4	8		21	0.6154	0.3810
Getting a ride	3						3			0	6	1.0000	0.0000
Total	17	45	6	2	495	17	38	32	13	0	665	**Overall accuracy**	0.9113

Acknowledgements The work reported in this paper was supported by the National Engineering Laboratory of China for Big Data System Computing Technology. The work reported in this paper was supported in part by the National Natural Science Foundation of China under grant no. 61836005 and no. 61672358, the Natural Science Foundation of Guangdong Province, China, under grant no. 2017A030310521, the Science and Technology Innovation Commission of Shenzhen, China, under grant no. JCYJ20160422151736824.

References

1. Huimin L, Li Y, Chen M, Kim H, Serikawa S (2018) Brain intelligence: go beyond artificial intelligence. Mobile Networks Appl 23:368–375
2. Huimin L, Li Y, Shengglin M, Wang D, Kim H, Serikawa S (2018) Motor anomaly detection for unmanned aerial vehicles using reinforcement learning. IEEE Internet Things J 5(4):2315–2322
3. Lu H, Wang D, Li Y, Li J, Li X, Kim H, Serikawa S, Humar I (2019) CONet: a cognitive ocean network. IEEE Wirel Commun, In Press
4. Serikawa S, Lu H (2014) Underwater image dehazing using joint trilateral filter. Comput Electr Eng 40(1):41–50
5. Huimin L, Li Y, Uemura T, Kim H, Serikawa S (2018) Low illumination underwater light field images reconstruction using deep convolutional neural networks. Future Gener Comput Syst 82:142–148
6. Hasan M, Roy-Chowdhury AK (2015) A continuous learning framework for activity recognition using deep hybrid feature models. IEEE Trans Multimedia 17(11):1909–1922
7. Samanta S, Chanda B (2014) Space-time facet model for human activity classification. IEEE Trans Multimedia 16(6):1525–1535
8. Wang F, Sun Z, Jiang Y-G, Ngo C-W (2014) Video event detection Using motion relativity and feature selection. IEEE Trans Multimedia 16(5):1303–1315

9. Cui P, Wang F, Sun L-F, Zhang J-W, Yang S-Q (2012) A matrix-based approach to unsupervised human action categorization. IEEE Trans Multimedia 14(1):102–110
10. Abbasnejad I, Sridharan S, Denman S, Fookes C, Lucey S (2016) complex event detection using joint max margin and semantic features. In: Proceedings of the international conference on digital image computing—techniques and applications. Gold Coast, QLD, Australia
11. Veeraraghavan H, Papanikolopoulos NP (2009) Learning to recognize video-based spatiotemporal events. IEEE Trans Intell Transp Syst 10(4):628–638
12. Kitani KM, Sato Y, Sugimoto A (2005) Deleted Interpolation using a hierarchical bayesian grammar network for recognizing human activity. In: Proceedings of the 2nd joint IEEE international workshop on VS-PETS, pp 239–246. Beijing
13. Shet VD, Harwood D, Davis LS (2005) VidMAP: video monitoring of activity with prolog. In: Proceedings of IEEE conference on advanced video and signal based surveillance (AVSS), pp 224–229
14. Song YC, Kautz H, Allen J, Swift M, Li Y, Luo J (2013) A Markov logic framework for recognizing complex events from multimodal data. In: Proceedings of the ACM on international conference on multimodal interaction (ICMI), pp 141–148
15. Liu L, Wang S, Hu B, Qiong Q, Wen J, Rosenblum DS (2018). Learning structures of interval-based Bayesian networks in probabilistic generative model for human complex activity recognition. Pattern Recogn 81:545–561
16. Song D, Kim C, Park S-K (2018) A multi-temporal framework for high-level activity analysis: violent event detection in visual surveillance. Inf Sci 447:83–103
17. Nawaz F, Janjua NK, Hussain OK (2019) PERCEPTUS: predictive complex event processing and reasoning for IoT-enabled supply chain. Knowl-Based Syst 180:133–146
18. Skarlatidis A, Artikis A, Filippou J, Paliouras G (2015) A probabilistic logic programming event calculus. Theory Pract Logic Program 15(2):213–245
19. Cavaliere D, Loia V, Saggese A, Senatore S, Vento M (2019) A human-like description of scene events for a proper UAV-based video content analysis. Knowl Based Syst 178:163–175
20. Azorin-Lopez J, Saval-Calvo M, Fuster-Guillo A, Garcia-Rodriguez J (2016) A novel prediction method for early recognition of global human behaviour in image sequences. Neural Process Lett 43:363–387
21. Castel C, Chaudron L, Tessier (1996). What is going on? A high-level interpretation of a sequence of images. In: Proceedings of the ECCV workshop on conceptual descriptions from images. Cambridge, U.K
22. Albanese M, Chellappa R, Moscato V, Antonio Picariello VS, Subrahmanian PT, Udrea O (2008) A constrained probabilistic Petri net framework for human activity detection in video. IEEE Trans Multimedia 10(6):982–996
23. Ghanem N, DeMenthon D, Doermann D, Davis L (2004) Representation and recognition of events in surveillance video using Petri nets. In: Proceedings of the 2004 IEEE computer society conference on computer vision and pattern recognition workshops (CVPRW'04)
24. Ghanem N (2007) Petri net models for event recognition in surveillance videos. Doctor thesis, University of Maryland
25. Lavee G, Rudzsky M, Rivlin E (2013) Propagating certainty in Petri nets for activity recognition. IEEE Trans Circuits Syst Video Technol 23(2):337–348
26. Lavee G, Borzin A, Rivlin E, Rudzsky M (2007) Building Petri nets from video event ontologies. In: Proceedings of the international symposium on visual computing, Part I, LNCS, vol 4841, pp 442–451
27. Lavee G, Rudzsky M, Rivlin E, Borzin A (2010) Video event modeling and recognition in generalized stochastic Petri nets. IEEE Trans Circuits Syst Video Technol 20(1):102–118
28. Borzin A, Rivlin E, Rudzsky M (2007) Surveillance event interpretation using generalized stochastic petri nets. In: Proceedings of the 8th international workshop on image analysis for multimedia interactive services (WIAMIS'07)
29. Ghrab NB, Boukhriss RR, Fendri E, Hammami M (2018) Abnormal high-level event recognition in parking lot. Adv Intell Syst Comput 736:389–398

30. Hamidun R, Kordi NE, Endut IR, Ishak SZ, Yusoff MFM (2015) Estimation of illegal crossing accident risk using stochastic petri nets. J Eng Sci Technol 10:81–93
31. Szwed P (2016) Modeling and recognition of video events with fuzzy semantic petri nets. Skulimowski AMJ, Kacprzyk J (eds) Knowledge, information and creativity support systems: recent trends, advances and solutions, pp 507–518
32. Szwed P (2014) Video event recognition with fuzzy semantic petri nets. Gruca A et al. (eds) Man-machine interactions, vol 3, pp 431–439
33. SanMiguel JC, Martínez JM (2012) A semantic-based probabilistic approach for real-time video event recognition. Comput Vis Image Underst 116:937–952
34. Liu L, Wang S, Guoxin S, Bin H, Peng Y, Xiong Q, Wen J (2017) A framework of mining semantic-based probabilistic event relations for complex activity recognition. Inf Sci 418–419:13–33
35. Kardas K, Cicekli N (2017) SVAS: surveillance video analysis system. Expert Syst Appl 89:343–361
36. Acampora G, Foggia P, Saggese A, Vento M (2015) A hierarchical neuro-fuzzy architecture for human behavior analysis. Inf Sci 310:130–148
37. Caruccio L, Polese G, Tortora G, Iannone D (2019) EDCAR: a knowledge representation framework to enhance automatic video surveillance. Expert Syst Appl 131:190–207
38. Murata T (1989) Petri nets: properties, analysis and applications. Proc IEEE 77(4):541–580
39. CAVIAR. http://groups.inf.ed.ac.uk/vision/CAVIAR/CAVIARDATA1/
40. VIRAT video dataset release 1.0. http://midas.kitware.com

Question Generalization in Conversation

Jianfeng Peng, Shenghua Zhong, and Peiqi Li

Abstract The dialogue response generation system is one of important topics in natural language processing, but the current system is difficult to produce human-like dialogues. The responses proposed by the chat-bot are only a passive answer or assentation, which does not arouse the desire of people to continue communicating. To address this challenge, in this paper, we propose a question generalization method with three types of question proposing schemes in different conversation patterns. A probability-triggered multiple conversion mechanism is used to control the system to actively propose different types of questions. In experiments, our proposed method demonstrates its effectiveness in dialogue response generalization on standard dataset. In addition, it achieves good performance in subjective conversational assessment.

Keywords Question generation · Probability-triggered · Dialogue response generalization system

1 Introduction

Proposing questions in dialogue has always been an indispensable way of learning human wisdom, as it can help assess the user's understanding of a piece of text [1]. With the popularity and development of dialogue generation system, chat-bots have been used more and more frequently in specific domain. Our day-to-day lives involve asking questions in conversations and dialogues to render a meaningful co-operative society, for example, Francisco et al. project a reminiscence-based social interaction tool [2] and Praveen et al. design a conversational system to aid new hires through their onboarding process [3]. But compared with human-like dialogues, the current responses obtained by automatic dialogue generation system are often boring [4] and tend to simple reply.

J. Peng · S. Zhong (✉) · P. Li
Shenzhen University, Shenzhen, China
e-mail: csshzhong@szu.edu.cn

© Springer Nature Switzerland AG 2021
H. Lu (ed.), *Artificial Intelligence and Robotics*,
Studies in Computational Intelligence 917,
https://doi.org/10.1007/978-3-030-56178-9_7

At present, in the production and optimization of dialogue generation system, the main idea of them is how to respond to the user's questions efficiently, or to give a variety of responses to the theme. However, in real life, the response is only one part of the daily conversation. Humans are almost subject to ambiguous, busy, or inconsistent mind in certain situations, question generation problem was raised to be solved and researched by many researchers over the last decade years. For this reason, we have made a preliminary attempt to actively propose the questions in the current dialogue generation system. This challenging attempt requires our model to do the following:

(1) Use different ways of asking questions in different conversations to avoid single questions pattern.
(2) The generated questions cannot be irrelevant to the original dialogue or far from the dialogue theme.

In recently years, deep learning has continually achieved breakthroughs in the field of natural language [4–6]. The data-hungry neural network based solutions require more data to meet the training of the model. Especially in some special natural language processing tasks, such as the question generation in dialogue domain. The construction of the dialogue question data set requires amount of manual annotation and evaluation. Therefore, creating newer datasets for specific domains or augmenting existing datasets with more data is a tedious, time-consuming and expensive process. To avoid this problem, in our system we use three artificial construction method that facilitates subsequent data construction and analysis.

In this paper, we propose three types of methods for manually throwing questions, and explains the role and principle of each method in detail. Besides, considering the different functions of different forms of questions in sentences, we explain the combination of different sentence patterns and the mechanism of the whole process. The methods we propose is as an external interface, which does not rely on a large number of current and relatively scarce questions corpus, to convert the response generated by a universal dialogue model into question sentence automatically. To construct a more accurate sentence-question conversion method, these three methods can be continuously optimized according to the researcher's preferences.

The rest of this paper is organized as follows. Section 2 briefly reviews the representative work on Sequence-to-Sequence (S2S). In Sect. 3, we introduce the details of using the machine framework to convert sentence into question in detail. In Sect. 4, our proposed method demonstrates its effectiveness in dialogue response generalization on benchmark dataset in the first experiment. In the second experiment, we get higher scores by our proposed model compared with the original sequences-to-sequences model. The results validate that the question generation model we propose can effectively improve the quality of the dialogue. The final conclusions are drawn in Sect. 5.

2 Related Work

In recent years, Deep Learning (DL) has surmounted many complex tasks in the field of natural language processing. It has reduced the difficulty of the research direction that was considered difficult to break through in the past.

In 2014, Sutskever et al. proposed a paper based on neural network machine translation [3]. This new paradigm was quickly applied to the automatic generation of chat responses, but it still could not reproduce the interaction of human communication. Subsequently, in order to solve a potential problem of encoder-decoder in the traditional model that information was compressed into vectors of fixed length and could not correspond to long sentences, bahdanau et al. proposed the attention mechanism [7] so that each word of each sentence in the new model had its own context. In 2015, Mukherjee used a large number of conversational data set training sequences into the sequence model and found that it was able to extract knowledge from a specific domain dataset and could perform simple common sense reasoning. Later, in view of the suggestion of attention, Yao et al. proposed using the three RNN structures of Encoder RNN, Intention RNN and Decoder RNN to train the dialogue system, and used the two mechanisms of attention and intention to improve the effect of multi-round dialogue [6]. Since the traditional seq2seq model usually had tended to generate some safe and normal replies, such as "*I don't know*" or "*I am ok*", Li et al. proposed a new objective function MMI to optimize seq2seq model [4], making the generated replies more diverse. In 2016, Li et al. proposed a method of using reinforcement learning to optimize the Seq2Seq parameters based on MMI [8] and achieved a better response effect. In 2018, Liu et al. added a distribution weight [9] to the objective function of Seq2Seq, which achieved an effect beyond MMI [8] and was able to remove safe and nutrient-free responses more effectively.

The work of the above scholars all focused on how to improve the quality of machine reply, but in the process of human-computer interaction, asking questions is also a very important step, and it can improve people's interest in dialogue so that the conversation will not be boring. Based on these cerebration, we shift the focus of our work to the question generation on dialogue generation model.

3 Method

In this paper, we propose three kinds of sentence-to-question transformation methods base on the English sentence structure and improve the beam search mechanism in the model for the direction of throwing more question instances. The principles and design of the methods will be detailed as follows. Section 3.1 describes the methods combination and the steps involved in building a conversation transform mechanism. Section 3.2 provides the details of the methods we have proposed. The definitions of different type questions in this paper are as follows:

- WH-type Question: English questions include *What*, *Who* and *Where* sentences.

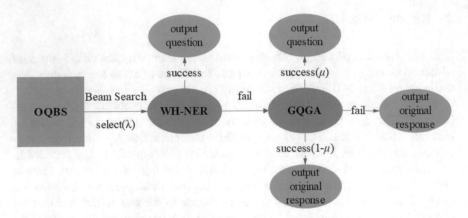

Fig. 1 Three methods probability trigger mechanism and transform output

- General Question (Yes-no question): The grammatical answer must be 'Yes' or 'No'. Its form tag is the inversion of the subject-verb word order. This "general" is contrary to the WH-type question.

3.1 Multiple Conversion Mechanism

Figure 1 shows the framework of the combination in our model. At the decoding layer, we first filter the results of model decoding and prioritize the **O**utput of the **Q**uestions from **B**eam **S**earch (**OQBS**) with the trigger probability of λ. The output sentence (not question) is then used as input to determine whether the sentence can be converted to a **WH**-type question by **N**ame **E**ntity **R**ecognizer (**WH-NER**). Since no entity in the sentence can be identified, the sentence will be entered into the next method and judge whether it can be successfully converted into a **G**eneral **Q**uestion by **G**rammatical **A**nalysis (**GQGA**) with the trigger probability of μ. In the case of determining that it can be successfully converted into a general question, we introduce an additional conversion technique to judge whether the existing **G**eneral **Q**uestion can be converted into a **WH**-type question (**GQ-WH**) with the trigger probability of $1 - \mu$.

3.2 Construction Method

As mentioned earlier, the sentence in the dialogue are different from the traditional Question-Asking [10–16] task dataset which are more targeted and normative as training data. Given this limitation, we propose three artificial construction method to generate different types of questions on the sequences-to-sequences model based

Table 1 First question word list

Symbol	Example
Modal-verb	Can, cannot, could, would, will…
Be verb	Is, are, were, was
Wh-determiner	Who, which, when, what
WH-pronoun	What, whatever
Possessive wh-pronoun	Whose
Practical verb "Do"	Do, does, did

chat-robot. The overall method consists of three parts: (i) Output existing Questions from Beam Search (**OQBS**). (ii) Declarative sentence to WH-type question (**WH-NER**), (iii) Declarative sentence to the General Question (**GQGA**).

- **Output existing Questions from Beam search**:

A. *Question identification*

In a normal conversation sentence, most of the question sentences can be judged from the first vocabulary combine with the definition of question at the beginning of Sect. 3. In this paper, we define the sentences of the first words as Modal-verb, Be verb, Wh-determiner, WH-pronoun, Possessive wh-pronoun and Practical verb "Do" as a question. The first words of the determined questions are shown in Table 1.

B. *Question selection*

With the shoring up of the question word, a question can be selected for the convenience of the response candidates. At the decoding layer, we use beam search [17] to generate N-best list replies of the highest probability of response. In order to select the most probable question from the candidate with a beam size of n, we use the above *Question identification* way to determine the question. Let $H = (h_1, h_2, \ldots h_n)$ be the candidate reply of the input sentence S, and $O = (o_1, o_2, \ldots o_n)$ is the probability set corresponding to the sentence in H. Mathematically, for an input S, the new generated probability sequence output the maximum probability sentence is that:

$$R = \left(q P(\arg\max(O_i)|o_1, o_2, \ldots o_n, H)|A_{original} \right) \tag{1}$$

where q represents the probability of method triggering, and $A_{original}$ represents the original statement to be output. When the trigger probability is not effective, the original sentence output will be selected.

We model the candidate set according to the probability distribution as:

$$P = (O|S) = P(o_i|o_1, o_2, \ldots o_n, H) \tag{2}$$

The simulation function for extracting the question is:

$$u = Q_i(h_1, h_2, \ldots h_n) \quad i_{start} = n \tag{3}$$

Here i_{start} is the list subscript to start locating. Q_i is a function for finding a question, and the result of the search is assigned to:

$$\begin{cases} u = i & if\,i\,in(1 \leq i \leq n) \\ u = 0 & if\,i = null \end{cases} \tag{4}$$

In the original list of output, the O_n is the maximum output probability of the candidate set. The probability of the question found O_u is reset as followed:

$$O_u = O_n + 1 \tag{5}$$

- **Declarative sentence to WH-type question**:

A. *Named entity recognition*

Stanford NER [18] is known as the CRF Classifier (conditional random field classifier). It provides a general implementation of a (random sequence) linear chain conditional random field (CRF) sequence model. In the following, we use Stanford NER to identify **Organization**, **Person**, **Location** named entities in vocabulary, and then match **What**, **Who**, **Where** three types of question corresponds to each of them. Figure 2 shows the correspondence between named entities and question types.

B. *Interlaced combination*

The purpose of asking question is to get more unknown information. We design an Interlaced combination of asking question based on the acquired entity information.

Fig. 2 Correspondence between named entities and question types

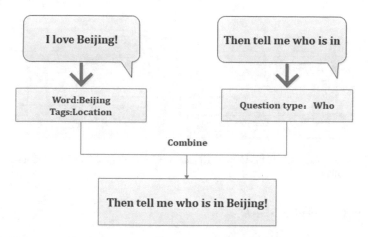

Fig. 3 Example of combination of named entity and question template

An example is shown in Fig. 3. If a sentence contains **Location** tag which associated with **Where** is identified, the algorithm will randomly select one from **What** or **Who** question templates and extract the specific words of the **Location** tag. It is combined as a key word in the question to form a brand new question. When two or more entities are identified in the sentence, we choose to directly ask the content of the entity tag without interlacing or cast the original statement.

- **Declarative sentence to the General Question**:

According to the basic structure unit of the general question which composed of auxiliary verb, subject and predicate verb, the sentence S_t is first divided into tokens $T_i = (T_1, T_2, T_3 \ldots T_i)$, where T_i refer to one word of sentence S_t, and the tagger and position information of each token are recorded by the python's NLTK Parser. The first auxiliary verb A_{verb} or modal verb M_{verb} found in the sentence will be advanced to the front of the subject word T_{sub}, where $S_t = (A_{verb}, T_{sub}, \ldots T_i | M_{verb}, T_{sub}, \ldots T_i)$. If the declarative sentence does not contain auxiliary verbs or modal verbs, add '*do (does, did)*' to reform a new general question before the subject word $S_t = (Do, T_{sub}, \ldots T_i)$. It is worth noting that, in order to ensure the grammatical correctness, the subject word and object word in the sentence also needs to be transformed where $S_t = (Do, T_{sub-transfrom}, \ldots T_i)$. for example, '*I*' needs to be transformed into '*you*' (Fig. 4).

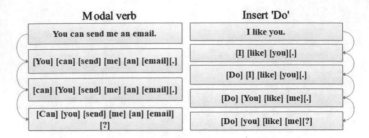

Fig. 4 Declarative sentence to the General Question example of modal verb being found and insert 'Do' in the sentence

4 Experiments and Results

In this section, we first describe the experimental setting we utilize for the evaluations. Secondly, we compare the question generation results obtained by our method with the baseline on Opensubtitles[1] dataset. The experimental result show that our method can not only increase the richness of the responses generated by the original model, but also effectively increase the number of dialogue rounds.

4.1 Experimental Setting

Datasets: The dialogue generation requires high-quality initial inputs fed to the agent. We conduct experiments on Opensubtitles[1], which contains a large number of movie subtitle data sets and is one of benchmark data set in dialogue generation. The advantage of using OpenSubtitles is that its conversation is a continuous conversation in a scene. We randomly select 2.5 million pairs for training, 1000 instances for validation and 500 instances for testing.

Baseline Dialogue Generation (NMT-Keras): Our baseline conversation generation model is Google Neural Machine Translation with Keras[2] (Theano and Tensorflow). This is a versatile translation model composed of Attentional recurrent neural network NMT model and Transformer NMT model.

Dialogue Generation with Question Generation (NMT-Q): Based on the original Neural Machine Translation with Keras, we propose our new dialogue generation system with question generation at the decoding layer and output layer.

Parameter setting: In the decoding layer, we use beam search with a beam size 10 to generate a response to a given input message. For the NMT-Q model, we first generate 10-best lists and then for the sentence that is not the question generated by

---

[1] https://www.opensubtitles.org/en/search/subs.

[2] https://github.com/lvapeab/nmt-keras.

the highest generation probability, we assign the probability of the first question had found to the maximum generation probability plus one in the candidate result. The probability of setting **OQBS** is 0.3. In the output layer, the probability setting for **WH-NER** is 0.3, If the conversion in **WH-NER** is failed, **GQGA** is triggered with a probability of 0.5, otherwise the original sentence will be output.

4.2   Automatic Evaluation

Evaluating dialogue systems is difficult. Metrics such as **BLEU-1** and perplexity have been widely used for dialogue quality evaluation. But it is widely debated how well these automatic metrics are correlated with true response quality. Since the goal of the proposed method is not to predict the highest probability response, but rather the long-term success of the dialogue, we do not regard BLEU-1 as the main reference standard.

We find the NMT-Q model perform worse on BLEU-1 score. On a sample of 500 conversational pairs, single reference BLEU-1 scores for NMT-Keras model and NMT-Q model are respectively 0.365 and 0.318. BLEU-1 is highly correlated with perplexity in generation tasks.

Evaluation metrics: In this experiment, two commonly metrics are used for evaluation:

- **Dialogue-Turns**: The first metric we use is that the number of dialogue rounds. Since the sequences-to-sequences model often generates repeated replies, we consider that for a pair of conversation groups, taking the shortest first N common subsequences and compare words one by one, if the same degree is greater than 85%, the dialogue ends.
- **Unigrams and Bigrams**: We report degree of diversity by calculating the number of distinct unigrams and bigrams in generated responses. This value is scaled according to the total number of tokens generated so that it can be used to support long sentences as described by Li et al. [8]. Thus the whole result metric is a type token ratio for unigrams and bigrams.

We compare to the base line model (NMT-Keras) and a model with three methods add (NMT-Q). To reduce the risk of circular dialogues, we limit the number of simulated turns to be less than 6. Directly applying 500 sentence as input train on OpenSubtitles leads to better performance with average number of dialogue turns 2.388. In the experiment, we still take a conservative attempt to set the trigger probability of all methods to below 0.3. In the case of low probability, the number of rounds of dialogue is indeed increased by constructing questions and increasing the occurrence probability of questions. The results are shown in Table 2.

When the NMT-Keras model generates a poor response, e.g. "*I don't know what to do with it.*", comparing the reply of NMT-Keras and NMT-Q model on the both sides of the Table 3, the proposed NMT-Q model can still generate better questions.

Table 2 The average number of dialogue turns from original NMT model and proposed question generation model

Model	Average number of dialogue turns
NMT-Keras	1.918
NMT-Q	2.388

Table 3 The NMT-Keras and NMT-Keras dialogue results and the NMT-Keras and NMT-Q dialogue results are shown separately

NMT-Keras (A) and NMT-Keras (B)	NMT-Keras (A) and NMT-Q (B)
INPUT: their reckless disregard in cairo brought on this violence today!	INPUT: their reckless disregard in cairo brought on this violence today!
A: I don't know what to do with it	A: I don't know what to do with it
B: I don't know, but I don't know	B: You know what?
A: I don't know, but I don't know	A: what you're talking about
B: I don't know, but I don't know	B: You know what I mean?
A: I don't know, but I don't know	A: Do you know what I can do?

It is noteworthy that at the starting with the second sentence thrown by the robot, if the question is generated, the subsequent sentence will change follow through. Contrast to the original model, which produces a large number of highly repetitive responses in the conversation, our approach solves this drawback to some extent.

Next, we report the Diversity scores of the two models, including NMT-Q and NMT-Keras. From Fig. 5, our proposed model achieves Unigram 0.028 and Bigram 0.103, both surpassing the base line model. The results demonstrate that using the proposed interface methods and the designed beam search algorithm, the overall

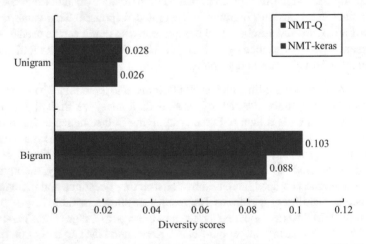

Fig. 5 Diversity scores (type-token ratios) for the standard NMT-Keras model and NMT-Q model

increase in sentence diversity has resulted in more versatile responses from machines. This situation also reflects from the side that the Beam Search mechanism that directly uses the maximum likelihood estimation is not conducive to solving specific problems in natural language processing, but only relies to the most common responses in the corpus.

5 Conclusion and Future Work

In this paper, we propose two types of question transformers and a search mechanism that can improve the beam search to increase the number of questions to a standard sequences-to-sequences model, aiming to different conversion contexts. At the same time, we use the probability-triggered interleaving combination mechanism to control the chat-bot to actively and reasonably propose different types of questions. Finally, we use the evaluation metrics of dialogue response generation system to prove that our proposed method can effectively improve the number of rounds of dialogue and the diversity of responses. The method we propose can be applied to any dialog model as an external interface. However, the current chat bots can't reach the level of human intelligence, and simply using the probabilistic trigger combination mechanism can sometimes be too rigid. In the future work, we will use reinforcement learning to optimize the probability trigger mechanism and the system can automatically determine the conversion context in a dialogue, which will propose questions more intelligently.

References

1. Kumar V, Boorla K, Meena Y, Ramakrishnan G, Li YF (2018) Automating reading comprehension by generating question and answer Pairs. In: Pacific-Asia conference on knowledge discovery and data mining. Springer, Cham, pp 335–348
2. Ibarra F, Baez M, Fiore F, Casati F (2017) Stimulating conversations in residential care through technology-mediated reminiscence. In: IFIP conference on human-computer interaction. Springer, Cham, pp 62–71
3. Chandar P, Khazaeni Y, Davis M, Muller M, Crasso M, Liao QV, Geyer W (2017) Leveraging conversational systems to assists new hires during onboarding. In: IFIP conference on human-computer interaction. Springer, Cham, pp 381–391
4. Li J, Galley M, Brockett C, Gao J, Dolan B (2015) A diversity-promoting objective function for neural conversation models. arXiv preprint arXiv:1510.03055
5. Wu B, Wang B, Xue H (2016) Ranking responses oriented to conversational relevance in chatbots. In: Proceedings of COLING 2016, the 26th international conference on computational linguistics: technical papers, pp 652–662
6. Yao K, Zweig G, Peng B (2015) Attention with intention for a neural network conversation model. arXiv preprint arXiv:1510.08565
7. Bahdanau D, Cho K, Bengio Y (2014) Neural machine translation by jointly learning to align and translate. arXiv preprint arXiv:1409.0473

8. Li J, Monroe W, Ritter A, Galley M, Gao J, Jurafsky D (2016) Deep reinforcement learning for dialogue generation. arXiv preprint arXiv:1606.01541
9. Liu Y, Bi W, Gao J, Liu X, Yao J, Shi S (2018) Towards less generic responses in neural conversation models: a statistical re-weighting method. In: Proceedings of the 2018 conference on empirical methods in natural language processing, pp 2769–2774
10. Li Y, Duan N, Zhou B et al (2018) Visual question generation as dual task of visual question answering. In: Proceedings of the IEEE conference on computer vision and pattern recognition, pp 6116–6124
11. Jain U, Zhang Z, Schwing AG Creativity: generating diverse questions using variational autoencoders. In: Computer vision and pattern recognition. IEEE, pp 6485–6494
12. Zhang J, Wu Q, Shen C et al (2017) Asking the difficult questions: goal-oriented visual question generation via intermediate rewards. arXiv:1711.07614
13. Du X, Shao J, Cardie C (2017) Learning to ask: neural question generation for reading comprehension. arXiv:1705.00106
14. Nan D, Duyu T, Peng C, Ming Z (2017) Question generation for question answering. In: EMNLP. Association for computational linguistics, pp 866–874
15. Liang P, Yang Y, Ji Y, Lu H, Shen HT (2019) Coarse to fine: improving VQA with cascaded-answering model. IEEE Trans Knowl Data Eng
16. Xu X, Lu H, Song J, Yang Y, Shen HT, Li X (2019) Ternary adversarial networks with self-supervision for zero-shot cross-modal retrieval. IEEE Trans Cybern (in Press)
17. Li J, Monroe W, Jurafsky D (2016) A simple, fast diverse decoding algorithm for neural generation. arXiv preprint arXiv:1611.08562
18. StanfordNER (2018) https://nlp.stanford.edu/software/CRF-NER.shtml. Last accessed 16 Oct 2018

Optimal Scheduling of IoT Tasks in Cloud-Fog Computing Networks

Zhiming He, Qiang Zhao, Haoran Mei, and Limei Peng

Abstract The huge volume of IoT data generated by emerging IoT end devices have triggered the prosperous development of Fog computing in the past years, mainly due to their real-time requirements. Fog computing aims at forming the idle edge devices that are in the vicinity of IoT end devices as instantaneous small-scale Fog networks (Fogs), so as to provide one-hop services to satisfy the real-time requirement. Since Fogs may consist of only wireless nodes, only wired nodes or both of them, it is significant to map IoT tasks with diverse QoS requirements to appropriate types of Fogs, in order to optimize the overall Fog performance in terms of the OPEX cost and transmission latency. Regarding this, we propose an integer linear programming (ILP) model to optimally map the IoT tasks to different Fogs and/or Cloud, taking into consideration of the task mobility and real-time requirements. Numerical results show that the real-time and mobility requirements have significant impact on the OPEX cost of the integrated Cloud-Fog (iCloudFog) framework.

Keywords Fog computing · Cloud computing · IoT · Real-time · Mobility

Z. He · Q. Zhao · H. Mei · L. Peng (✉)
School of Computer Science and Engineering, Kyungpook National University, Daegu 41566, South Korea
e-mail: auroraplm@knu.ac.kr

Z. He
e-mail: hezhimingabc@knu.ac.kr

Q. Zhao
e-mail: zhaoqiang@knu.ac.kr

H. Mei
e-mail: meihaoran@knu.ac.kr

© Springer Nature Switzerland AG 2021
H. Lu (ed.), *Artificial Intelligence and Robotics*,
Studies in Computational Intelligence 917,
https://doi.org/10.1007/978-3-030-56178-9_8

103

1   Introduction

The current datacenters in the Cloud are fixed and distant from the IoT end devices and thus are short of providing real-time services and supporting task mobility, when confronting the emerging IoT tasks with features of velocity, volume and variety. Regarding this, Fog computing was coined with the motivation of provisioning services by edge devices in their vicinity, so as to minimize the transmission latency for real-time IoT tasks. In the past years, great progresses have been made in Fog computing in various aspects, such as Fog planning [1–5], energy consumption, Fog applications in various fields, emerging Fog computing with 5G [6], etc.

Specifically, authors in [7] proposed an integrated Cloud-Fog computing framework and described the major challenges and solutions. In [8, 9], the authors proposed an optimized Fog planning model aiming at minimizing the overall network delay and the number of tasks sent to the Cloud taking into account of the different attributes of end devices, Fog nodes and links, and optimal installation location of Fog nodes.

Authors in [10] tried to apply Fog computing to the traffic system by compressing high frame-rate video streams at Fog nodes. In [11], the authors proposed to emerge the multi-access Edge computing with 5G networks, which was said to be able to improve the QoS and utilize the mobile backhaul and core networks more efficiently. The authors in [12] provided a theoretical analysis to optimize the power consumption coming from the content caching and dissemination in hot spots, fronthaul links, and rural areas. The use of backhaul/fronthaul links was traded off against the efficiency of content distribution. In [13], the authors designed an architecture that addressed some of the major challenges for the convergence of Network Function Virtualization (NFV), 5G/Mobile Edge Computing (MEC), IoT and Fog computing.

Most of the existing literatures focused on Fog planning, energy consumption, application of Fog computing, convergence of Fog and advanced 5G technologies, etc. Very few of them has addressed the important issue of mapping the huge IoT tasks to the integrated Cloud and Fog networks (iCloudFog). Regarding this, we set our goal as optimally mapping IoT tasks to dynamic Fogs and/or Cloud. Specifically, we propose an integer linear programming (ILP) model with the objective of minimizing the overall OPEX cost and transmission delay. In the proposed model, we take into account of both the diverse QoS requirements of IoT tasks, such as real-time and mobility requirements, and the network attributes, such as the resource availability status of different Fogs.

The rest of this paper is structured as follows. In Sect. 2, the integrated Cloud-Fog (iCloudFog) architecture together with the characteristics of IoT tasks are introduced. In Sect. 3, we introduce the proposed ILP model in details. The numerical results of the proposed ILP model based on AMPL are given and analyzed in Sect. 4. Section 5 concludes this paper.

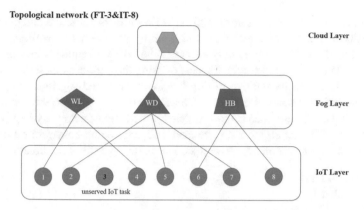

Fig. 1 iCloudFog framework. FT-3: three Fog types; IT-8: eight IoT tasks; WL: wireless Fog; WD: wired Fog; HB: hybrid Fog

2 ICloudFog and Fog Networks

In this section, we introduce the integrated Cloud-Fog (iCloudFog) framework as shown in Fig. 1. It consists of three layers, say the Cloud, the Fog and the IoT end layers from top to bottom.

For the Fog layer, since we assume that a Fog may consist of only wireless edge devices, only wired edge devices, or both of wireless and wired edge devices, we classify the Fogs into three types, named as wireless Fog (WL), wired Fog (WD), and hybrid Fog (HB), respectively [7]. Note that the three types of Fogs differ from each other in various aspects, such as the computing/storage capability, the competence of handling mobile tasks, etc., and thus are appropriate for IoT tasks with different QoS requirements.

For IoT tasks generated by IoT end devices, we mainly consider the requirements of real-time and mobility. Lightweight delay-sensitive IoT tasks and/or mobile IoT tasks are more likely to be handled by wireless Fogs or hybrid Fogs. Heavyweight IoT tasks with no requirement on real-time and mobility can be forwarded to wired Fogs or hybrid Fogs. If the Fog resources are saturated, the heavyweight IoT tasks can also be forwarded to the Cloud for processing, albeit at a slightly higher cost.

3 ILP Model of IoT Task Scheduling

Based on the above iCloudFog framework, we develop an integer linear programming (ILP) model in this section, with the objective of minimizing the operating expense (OPEX) and maximizing the total number of IoT tasks that are successfully served.

As introduced previously, we consider three types of Fogs, say wireless Fog, wired Fog, and hybrid wired/wireless Fog. In addition, the IoT tasks are featured by their requirements on real-time and/or mobility. We assume that wireless and hybrid Fogs can handle IoT tasks with real-time and mobility requirements, while wired Fogs cannot. For the OPEX, we mainly consider the cost on requiring resources from different Fog types. For example, the cost of using the wireless Fog links is the most expensive, followed by the hybrid Fog links and the wired fog links. Suppose the wired and hybrid Fogs are directly connected with the Cloud in the top layer, the cost of using the Cloud links is more expensive than that of using any Fog types. For the proposed ILP model, the set, parameters, objectives and constraints are given as follows. Note that we consider two objectives under two different situations.

(1) Sets and Parameters
    See Table 1.
(2) Decision Variables

- $st_{f,i}$: Binary variable, which is one if IoT task i is successfully served by fog type f, i.e., $st_{f,i} = 1$; zero, vice versa, $f \in FT$ and $i \in IT$.

Table 1 Summary of parameters used in the ilp model

FT	Set of Fog types
IT	Set of IoT tasks
FTC_f	Maximum computing resource available of fog type $f, f \in FT$
FTS_f	Maximum storage resource available of fog type $f, f \in FT$
$FTRT_f$	Binary parameter. One indicates fog type f can support real-time task; zero, vice versa ; $f \in FT$
$FTMB_f$	Binary parameter. One indicates fog type f can support mobility task; zero, vice versa; $f \in FT$
$FT2C_f$	Binary parameter. One indicates fog type f can connect to Cloud; zero, vice versa; $f \in FT$
$FTTQ_f$	The maximum number of IoT tasks fog type f can handle, $f \in FT$
ITC_i	Total amount of computing resource required with IoT task $i, i \in IT$
ITS_i	Total amount of storage resource required with IoT task $i, i \in IT$
$ITRT_i$	Binary parameter. One indicates IoT task i is real-time tasks; zero, vice versa; $i \in IT$
$ITMB_i$	Binary parameter. One indicates IoT task i is mobility tasks; zero, vice versa; $i \in IT$
$OpEx\_FT2C_f$	OPEX when accessing Cloud via Fog type $f, f \in FT$
$OpEx\_IT2FT_f$	OPEX when Fog type f is selected to serve an IoT task, $f \in FT$
$st_{f,i}$	Binary variable, which is one if IoT task i is successfully served by fog type f, $st_{f,i} = 1$; and zero, vice versa, $f \in FT$ and $i \in IT$
$it2c_i$	Binary variable, which is one if IoT task i is handled by Cloud due to insufficient computing resources from Fogs; zero, vice versa, $i \in IT$
α	A weight value, used for weighted sum to solve multi-objective optimization

- $it2c_i$: Binary variable, which is one if IoT task i is handled by Cloud due to insufficient computing resource from Fogs, i.e., $it2c_i = 1$; zero, vice versa, $i \in IT$.

(3) Objective Functions

- Minimize (Total_OPEX):

$$\sum_{f,i}^{f \in FT, i \in IT} st_{f,i} * \mathrm{OpEx\_IT2FT}_f + \sum_{f,i}^{f \in FT, i \in IT} st_{f,i} * it2c_i * \mathrm{OpEx}_{FT2Cf} \qquad (1)$$

- Maximize (Success_Tasks):

$$\sum_{f,i}^{f \in FT, i \in IT} st_{f,i} \qquad (2)$$

Our Objective is to minimize the OPEX while maximizing the number of IoT tasks successfully served under the iCloudFog framework as shown in Eqs. (1) and (2). The first objective in (1), say Total_OPEX, aims at minimizing the OPEX due to using Fog links and Cloud links. The second objective in (2), say Success_Tasks, aims at maximizing the total number of IoT tasks that are successfully served.

- Weighted Sum Optimization

Instead of focusing on a single goal, we consider a weighted sum optimization objective as shown in Eq. (3), i.e., Total_Objective, aiming at minimizing the total OPEX due to using Fog and/or Cloud links and meanwhile maximizing the total number of IoT tasks that are successfully served.

Total_Objective:

$$\text{Minimize (Total\_OPEX} - \alpha * \text{Success\_Tasks)} \qquad (3)$$

(4) Constraints

$$\sum_{f}^{f \in FT} st_{f,i} \leq 1, \forall i \in IT \qquad (4)$$

$$st_{f,i} * \mathrm{ITRT}_i \leq \mathrm{FTRT}_f, \forall i \in IT \qquad (5)$$

$$st_{f,i} * it2c_i \leq (1 - \mathrm{ITRT}_i) * \mathrm{FT2C}_f, \forall i \in IT \qquad (6)$$

$$\sum_{i}^{i \in IT} st_{f,i} * \text{ITC}_i * (1 - it2c_i) \leq \text{FTC}_f, \forall j \in \text{FT} \tag{7}$$

$$\sum_{i}^{i \in IT} st_{f,i} * \text{ITS}_i * (1 - it2c_i) \leq \text{FTS}_f, \forall j \in \text{FT} \tag{8}$$

$$st_{f,i} * \text{ITM}B_i \leq \text{FTM}B_f, \forall i \in \text{IT} \tag{9}$$

$$\sum_{i}^{i \in IT} st_{f,i} * (1 - it2c_i) \leq \text{FTTQ}_f, \forall j \in FT \tag{10}$$

Constraint (4) ensures that any IoT task can only be served by one Fog type at most. Constraint (5) ensures that when an IoT task requires real-time service, the Fog type serving it must have the ability to handle real-time tasks. Constraint (6) ensures that if an IoT task is a real-time task, it can only be served by Fogs but cannot be forwarded to the cloud. Vice versa, if it is not a real-time task, it can be uploaded to the Cloud for processing. Constraints (7) and (8) ensure that for any Fog type, the sum of demands on computing/storage resource of all IoT tasks served by it cannot exceed its total amount of computing/storage resource. Constraint (9) ensures that for any IoT task with mobility requirement, it can only be served by Fogs which can provide mobility. For IoT tasks with no mobility requirement, they can be served by any of the Fogs or Cloud. Constraint (10) ensures that for any Fog type, the total number of IoT tasks that are successfully served must be no larger than the total number of IoT task demands.

4   Numerical Evaluation

(1)   Network Environment

We consider a topology as shown in Fig. 1. More specifically, we assume the computing resource of the Cloud is sufficient. We consider three types of Fogs in the middle layer, say one wireless Fog (WL), one wired Fog (WD), and one hybrid Fog (HB). The wireless and hybrid Fogs can serve IoT tasks with real-time and mobility requirements. The wired and hybrid Fogs can forward the IoT tasks to the Cloud. The total number of IoT task demands are set to be 32. The mobility and real-time requirement of all the 32 IoT tasks are generated randomly, ranging from 5 to 20. The cost of using Fog and Cloud links are set as follows. The costs of using the wireless Fogs, hybrid Fogs and wired Fogs are set as 5, 15, and 30, respectively. In addition, the OPEX cost of using wired Fogs to Cloud and the OPEX cost of using hybrid Fogs to Cloud are 30 and 70 respectively. With the above parameters, we run the proposed ILP models using AMPL (Fig. 2).

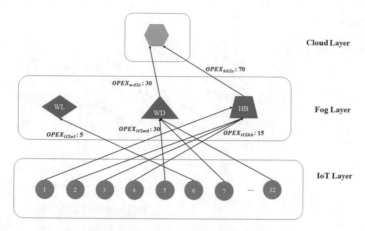

Fig. 2 Value Setting of ILP Model. $OPEX_{wd2c}/OPEX_{hb2c}$: the OPEX cost of traversing wired/hybrid Fogs to Cloud link. $OPEX_{it2wl}/OPEX_{it2wd}/OPEX_{it2hb}$: the OPEX cost of provisioning IoT tasks via wireless/wired/hybrid Fogs link

(2) Numerical Results

Figures 3 and 4 show the total OPEX cost under the objective 1 in Eq. 1 by assuming all the 32 IoT tasks have been served successfully. In Fig. 3, we assume two cases differing in the numbers of IoT tasks with mobility (MB), i.e., MB = 0 and 5, respectively. We can observe that the total OPEX costs of both cases increase with increasing number of IoT tasks requiring real-time (RT) services. Similarly, in Fig. 4, we assume two cases differing in the number of IoT tasks with real-time requirement (RT), i.e., RT = 0 and 5, respectively. We can observe that the total OPEX costs of both cases increase with the increasing number of IoT tasks requiring mobility services.

Fig. 3 Total OPEX cost under objective 1 in Eq. 1. MB: # of mobile IoT tasks

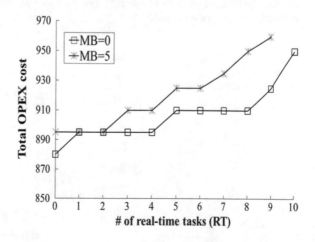

Fig. 4 Total OPEX cost
under objective 1 in Eq. 1.
RT: # of real-time IoT tasks

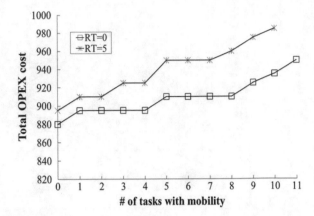

Figures 5, 6, and 7 show the numerical results under the weighted objective 3 in
Eq. 3 which aims at minimizing the OPEX cost and meanwhile maximizing the total
number of IoT tasks successfully served. Figure 5 shows the impact of the number
of IoT tasks successfully served on the total OPEX cost. We can observe that the

Scope of α	Sum_ST	Total_OpEx
$(110, +\infty)$	32	1085
$(85, 110]$	31	975
$[45, 85]$	26	550
$(35, 45)$	25	505
$(30, 35]$	22	400
$(15, 30]$	14	160
$(5, 15]$	5	25
$(-\infty, 5]$	0	0

Fig. 5 a Total OPEX cost under objective 3 (Eq. 3). MB: # of mobile IoT tasks. **b** The range of α
and the impact of α on the objectives

Fig. 6 Total OPEX cost
under objective 3 in Eq. 3.
MB: # of mobile IoT tasks

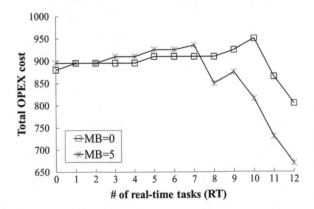

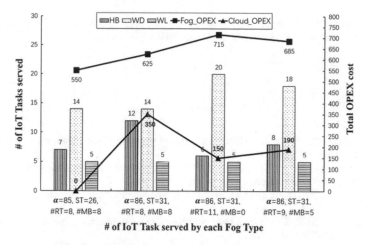

Fig. 7 Total OPEX versus # of IoT tasks server under objective 3 in Eq. 3

total OPEX increases with increasing number of successfully served IoT tasks which is reasonable. Figure 6a shows the total OPEX cost with the increasing RTs under different MBs, say MB = 0 and 5, respectively, when the weight value of α is set to 86. We can observe that when the number of RT reaches 7 and 10 for MB = 5 and 0, respectively, the number of total OPEX cost decreases dynamically. This is attributed to the fact that the number of IoT tasks successfully served reduces dynamically because of lack of resource in Fogs that can support real-time and/or mobility services. Figure 7 shows the OPEX cost and the total number of IoT tasks that are successfully served by different Fog types under different RT and MB values. We can observe that the WD Fogs are used the most frequently, followed by the HB Fogs and the WL Fogs.

5 Conclusion

In this paper, we proposed an Integer Linear Programming (ILP) model to optimally map the IoT tasks with different Fogs and Cloud, taking the attribute of mobility and real-time requirement of IoT tasks into consideration. The objective of the proposed ILP model was to minimize the OPEX cost of the Fog/Cloud and maximize the number of successfully served IoT tasks. Numerical results have shown that the QoS requirements of IoT tasks such as real-time service, mobility service, etc., impact the iCloudFog performance significantly.

Acknowledgements This study was supported in part by the BK21 Plus project (SW Human Resource Development Program for Supporting Smart Life) funded by the Ministry of Education, School of Computer Science and Engineering, Kyungpook National University, Korea

(21A20131600005) and in part by the National Research Foundation of Korea (NRF) Grant funded by the Korean government (Grand No. 2018R1D1A1B07051118).

References

1. Zhang D et al (2019) Model and algorithms for the planning of fog computing networks. IEEE Internet Things J 6(2):3873–3884
2. Yousefpour A et al (2019) FogPlan: a lightweight QoS-aware dynamic fog service provision-in-g Framework. IEEE Internet Things J 6(3):5080–5096
3. Li Y, Jiang Y, Tian D, Hu L, Lu H, Yuan Z (2019) AI-enabled emotion communication. IEEE Netw 33(6):15–21
4. Lu H, Liu G, Li Y, Kim H, Serikawa S (2019) Cognitive Internet of vehicles for automatic driving. IEEE Netw 33(3):65–73
5. Lu H, Wang D, Li Y, Li J, Li X, Kim H, Serikawa S, Humar I (2019) CONet: a cognitive ocean network. IEEE Wirel Commun 26(3):90–96
6. Sodhro A et al (2018) 5G-based transmission power control mechanism in Fog computing for internet of things devices. Sustainability 10(4):1258
7. Peng L et al (2018) Toward integrated Cloud-Fog networks for efficient IoT provisioning: Key challenges and solutions. Future Gener Comput Syst 88:606–613
8. Haider F (2018) On the planning and design problem of fog networks. Carleton University
9. Deng R et al. (2015) Towards power consumption-delay tradeoff by workload allocation in cloud-fog computing. In: IEEE international conference on communications (ICC), London
10. Liu J et al (2018) Secure intelligent traffic light control using fog computing. Future Gener Comput Syst 78(2):817–824
11. Taleb T et al (2017) On multi-access edge computing: a survey of the emerging 5G network edge cloud architecture and orchestration. IEEE Commun Surv Tutor 19(3):1657–1681
12. Lien S Y et al (2018) Energy-optimal edge content cache and dissemination: designs for practical network deployment. IEEE Commun Mag 56(5):88–93
13. Van Lingen F et al (2017) The unavoidable convergence of NFV, 5G, and fog: a model-driven approach to bridge cloud and edge. IEEE Commun Mag 55(8):28–35

Context-Aware Based Discriminative Siamese Neural Network for Face Verification

Qiang Zhou, Tao Lu, Yanduo Zhang, Zixiang Xiong, Hui Chen, and Yuntao Wu

Abstract Although face recognition and verification algorithms have made great success under controlled conditions in recent years. In real-world uncontrolled application scenarios, there is a fundamental challenge that how to guarantee the discriminative ability of feature from vary inputs for face verification task. Aiming at this problem, we proposed a context-aware based discriminative siamese neural network for face verification. In fact, the structure of facial image are more stable rather than hairstyle change and wearing jewelry. Firstly we use a context-aware module to anchor facial structure information by filtering out irrelevant information. For improved discrimination, we develop a siamese network including two symmetrical branch subnetworks to learn discriminative feature by labeled triad training data. The experimental results on LFW face dataset outperform some state-of-the-art face verification methods.

Q. Zhou · T. Lu (✉)
Hubei Key Laboratory of Intelligent Robot, Wuhan Institute of Technology, Wuhan, China
e-mail: lutxyl@gmail.com

Q. Zhou
e-mail: zq315653752@icloud.com

Y. Zhang · Y. Wu
School of Computer Science and Engineering, Wuhan Institute of Technology, Wuhan, China
e-mail: zhangyanduo@hotmail.com

Y. Wu
e-mail: ytwu@sina.com

Z. Xiong
Department of Electrical and Computer Engineering, Texas A&M University,
College Station, USA
e-mail: zx@ece.tamu.edu

H. Chen
Simshine-WIT Joint Laboratory of Perceptual Intelligence,
Wuhan Institute of Technology, Wuhan, China
e-mail: hui.chen@simshine.cn

© Springer Nature Switzerland AG 2021
H. Lu (ed.), *Artificial Intelligence and Robotics*,
Studies in Computational Intelligence 917,
https://doi.org/10.1007/978-3-030-56178-9_9

1 Introduction

In recent years, face verification as an important biometric authentication manner, is widely used in many fields such as military, security, and finance. Indeed, face verification has long history in the field of computer vision [22]. In the early stages, traditional face verification methods mainly utilize the geometrical structure of the face to verify the feature points by organs and topological relationship between registered images and testing ones [2]. Intuitively, some special-designed hand-craft features such as HOG [6], LBP [1] had performed well under controlled configurations. Equipped with some sophisticated classifiers, such as Support Vector Machine (SVM) [13], boosting [16], and Sparse Representation classifier [23] have successively explored face verification. Due to the limitation of hand-craft learned features, above methods improve the accuracy of the LFW [10] benchmark to about 95% [3]. How to design the discriminative features and robust performance against unconstrained facial variations is still a challenge. Since AlexNet won the ImageNet competition using deep neural network [12], deep-learning based paradigm offers an end-to-end learning tool without designing hand-craft features.

Schroff et al. [19] proposed an unified embedding method for learning features, called "Facenet". A triplet loss by l_2 normalization of embedded features was used to boost its discriminative ability. Wu et al. [24] used a lightweight convolutional neural network (LCNN) to accelerate recognition effectiveness. Deng et al. [7] incorporated angular margin and cosine margin into loss function to enhance the discriminative ability of features. Chen et al. [4] proposed depth-wise separable convolutional neural network for efficient mobile vision applications (Mobilefacenet). Although above mentioned methods yield good performance from the point of enhancing discriminative power of loss function. In fact, the discriminative information from training label has good potential insight for further improving discriminative ability of the learned features.

In the case of real world applications, lighting condition [20], noise, low-resolution and low-quality image [15], pose, expression or even guise make great challenges for face verification task. One basic question is how to learn discriminative features. Inspired by attention mechanism [14], in this paper, we propose a context-aware module to process face images, remove insignificant background information, and focus on face structure information for face verification. Considering the discriminative information from face label, the Siamese network is widely used for enhancing discriminative power of features. Chopra et al. [5] learned discriminative metric by differing the similarity metric to be as large as possible for same person face pairs, and as small as possible for pairs from different persons. Some representative Siamese networks such as [9, 11, 18] enhanced the power of features discriminatively. However, those methods only use simple two branch structure to differ the features which limits their performance. Besides, they simply use off-the-shelf face detection algorithms to locate the face image involving some unrelated facial parts such as hair or background. In this paper, we develop a disrciminative siamese network with context-aware model to focus on the structure discriminative information from facial images. First, we use context-aware model to extract structure information with filtering out unrelated information. Then, we use the improved Siamese network as the backbone and the cross entropy loss function to guide the training. Our method achieves better performance with discriminative features for face verification than some state-of-the-art methods. To summarize, the contributions of this paper are highlighted as follows:

(1) We firstly propose a context-aware Siamese network for face verification. The proposed context-aware model can adaptively focus on core region of facial images with removing irrelevant features. The landmarks detecting process automatically aligns facial image into fixed grid.

(2) We proposed to use a improved discriminative Siamese network to learn sematic robust features. Furthermore, we use a metric to measure the discriminative ability of features which can be used in other related vision based tasks.

2 Context-Aware Based Discriminative Siamese Neural Network (CDSN)

The proposed CDSN is consisted by four parts: Input part, context-aware module, enhanced Siamese network, and output part. The architecture of CDSN is shown in Fig. 1.

2.1 Context-Aware Module

In this paper, context-aware module is used to remove the complex background, hairstyle and other information with focusing on main organs on facial images. We simply use face landmark detection algorithm [25] to locate the key points of facial image. Based on a basic observation that a rectangle is suitable to figure out the

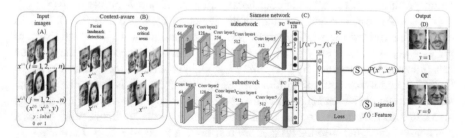

Fig. 1 Architecture of CDSN. **a** Input images; **b** context-aware module, its role is to find the most important information area of the input face image; **c** Enhanced Siamese neural network, The structures of two branches of CDSN are the same, with sharing the same weight; **d** Output labels

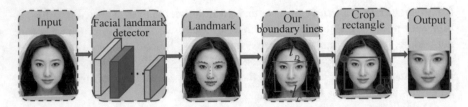

Fig. 2 Process of context-aware module. From left to right: Input facial image, landmark detector, the 68 key points of facial image, location of boundary lines from landmarks, the cropped rectangle according to the four edge points, output of facial image

main organs of facial image. We assume a minimum rectangle to crop out the face content. The process of context-aware module is shown in Fig. 2. Here, we define facial image samples $I_i{}_{i=1}^N \in R^{m \times n}$, i is the index of samples, N is the number of all samples, m and n indicate the height and width of the facial image. The purpose of context-aware module is to shrink the image I_i into main parts of facial image $X_i \in R^{u \times v}$, u and v is the height and width of main content of facial image. In this paper, we use three-steps to detect the content of facial images. First, we define face landmark detection function D as:

$$D(I_i) = \{(x_j, y_j)|1 \le j \le 68\}. \tag{1}$$

Second, from above 68 points, we need to find four points as the boundary lines. Here these four boundary points are defined as:

$$
\begin{aligned}
l_A &= \{x_a = min\{x_1, x_2, \dots, x_{68}\}, y_j\}, \\
l_B &= \{x_b = max\{x_1, x_2, \dots, x_{68}\}, y_j\}, \\
l_C &= \{x_j, y_c = min\{y_1, y_2, \dots, y_{68}\}\}, \\
l_D &= \{x_j, y_d = max\{y_1, y_2, \dots, y_{68}\}\}.
\end{aligned}
\tag{2}
$$

The intersection of the four boundary lines yields a rectangle. We use this rectangle's upper left corner point E and the lower right corner point F to represent this rectangle. It is easy to solve the coordinate values of E and F points by $E(x_a, y_c)$, $F(x_c, y_d)$. At last, considering inaccurate detection of key points, we intend to enlarge the yielded rectangle slightly. We add a constant w to adjust the rectangle as $E(x_a - w, y_c - w) F(x_c + w, y_d + w)$. In our implementation, the cropped image size L is determined by $L = max(d_v, d_h)$. Where $d_v v$ is the distance between the upper and lower landmark points, $d_v = y_d - y_c + 2w$, d_h is the horizontal distance between the left and right landmark points, $d_h = x_c - x_a + 2w$. Once the crop size L is determined, we crop the facial region center on the landmark point of nose. For convenience, the cropped images are resized to a fixed size 72×72 pixels.

2.2 Enhanced Siamese Neural Network

When content of facial image X_i is ready, we define a triplet data $(x^{(i)}, x^{(j)}, y)$, where $x^{(i)} \in X_i$ and $x^{(j)} \in X_j$ represent two images from training samples, where y is label. If image $x^{(i)}$ and $x^{(j)}$ are from one same person, then label $y = 1$, otherwise $y = 0$. Different with shallow Siamese network [5], we propose a deeper Siamese neural network to enhance its discrimination. Therefore, we use the cross entropy loss function as the loss function of the network, the overall loss function is defined as:

$$loss = \sum_{i=1}^{N}(-(y \log \hat{y}(1 - y) \log(1 - \hat{y}))). \tag{3}$$

where N represents the number of training sample pairs, y represents the real label, and $\hat{y}$ represents the predicted label. Since we paired the trained databases in pairwise with assigned labels. Face verification becomes a classification problem. Specifically, given a pair of face images as input, thus the feature vectors generated by the Siamese network are input to the same Sigmoid unit. Here the output is $y = 1$ means that these two images are recognized as the same person, otherwise $y = 0$ means different person. The Sigmoid unit is defined as:

$$\hat{y} = \sigma(\sum_{k-1}^{K} \omega_k \frac{(f(x^{(i)})_k - f(x^{(j)})_k)^2}{f(x^{(i)})_k - f(x^{(j)})_k} \mid b), \tag{4}$$

where $\frac{(f(x^{(i)})_k - f(x^{(j)})_k)^2}{f(x^{(i)})_k - f(x^{(j)})_k}$ is called χ Square similarity, $\hat{y}$ represents the predict label which valued $[0, 1]$. $f(x^{(i)})_k$ and $f(x^{(j)})_k$ represents the extracted k dimensional feature vector from image $x^{(i)}$ and $x^{(j)}$, ω_k indicates the initial fixed vector of the network, b indicates the initial fixed value of the network, and $\sigma_{(t)}$ defined as:

$$\sigma_{(t)} = \frac{1}{1+e^{-t}}, \tag{5}$$

thus the prediction $\hat{y}$ label is defined as:

$$\hat{y} = \frac{1}{1+\exp(\sum_{k=1}^{K} \omega_k \frac{(f(x^{(i)})_k - f(x^{(j)})_k)^2}{f(x^{(i)})_k - f(x^{(j)})_k} +b)}. \tag{6}$$

The value of $\hat{y}$ is $[0, 1]$. When the value of $\hat{y}$ is closer to 1, it means that the identity of the two faces is the same. When the value of $\hat{y}$ is closer to 0, the identity of the person in the two faces is different.

3   Experiments

3.1   Database Description

We conduct experiments using Microsoft's published aligned celebrity database (MS-Celeb-1M) [8]. This database contains about 79,800 different person, each with at least 50 different images. Here, we randomly select 50,000 different person which each contains 30 images. As we know, sample-pairs from same person with label value is 1, otherwise, sample-pair from different person with label 0. Here, the total number of samples-pairs is 1.5 million with half positive samples and half negative samples. In the testing phrase, we chose the LFW database (we remove 600 images that were unqualified when processed by context-aware module), and the real-world database (including 300 person, 10 different images per person) to testify the proposed method.

3.2   The Role of Context-Aware Module

In order to verify the validation of context-aware model, we test the CDSN with different settings of control parameter w. Firstly, we set $w = 0$ to visualize the distribution of facial images. 80 images (10 person with 8 different images) from LFW database are randomly selected to form the testing dataset. Images with original size and with the context-aware size are used to compared to show the role of context-aware module. We train two LCNN [24] models with above two settings and extract the testing samples with the trained module. As shown in Fig. 3, testing samples are mapped to 2-D vectors using T-SNE [17].

Obviously, context-aware module focus on the main organs of facial images, without the background, the context-aware module play an important role in clustering the samples into better discriminative distribution.

On the other hand, we adjust the control parameter w to testify the performance of context-aware module. As we know, the value of w controls the size of key area of facial image. Since point $E(x_a - w, y_c - w)$ and point $F(x_b + w, y_d + w)$ control the size of the cropped rectangle, it is specified that when $w = +\infty$, just select the original image. When $w > 0$, the rectangle will expand, otherwise $w < 0$ the rectangle will shrink. We use the cropped images to train and test CDSN. From Fig. 4, when $w = 3$, the face information contained in the cropped rectangle is the most related information for CDSN and it achieves its best performance. Moreover, too big w involves more unrelated information, but too small w may ignore some important information.

Mapping of LCNN descriptors in 2d via T-SNE

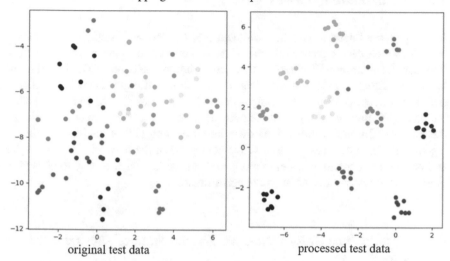

original test data   processed test data

Fig. 3 Visualization of facial features in 2-D vector space. Here, the different colors represent different person. Samples processed by the context-aware module has compact distribution of in features space. The context-aware module squeeze discriminative information for focusing on main organs of facial image

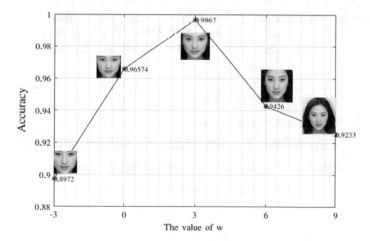

Fig. 4 The value of w control the performance of CDSN. Obviously, a suitable value gains most accurate information for verification task

3.3 Architecture of CDSN

In this part, we intend to test the architecture of CDSN. With stable context-aware module, we cascade enhanced Siamese neural network to figure out face verification task. Comparing with classic 5-layer convolutional network, CDSN used more layers for extracting discriminative features. As shown in Fig. 5, the results indicate that deeper network architecture outperforms shallow ones. The main reason is that increasing the number of layers of the network layer can increase the nonlinearity of the network. This makes the network owns better learning ability and further enhancing the discriminative feature ability. Inevitably, when the network level is deeper and deeper, the deep network gradient is very unstable, which is easy to cause the gradient to disappear, results in worse performance.

3.4 Performance Comparison with Some State-of-the-Art Algorithms

Here, we use CDSN to testify verification performance with some representative face verification algorithms. As shown in Table 1, Deepface [21], LCNN [24], Facenet [19], Mobilefacenet [4], Arcface [7] as the benchmarks. As we known, Arcface is the state-of-the-art verification algorithm, and MobilefaceNet is the state-of-the-art algorithm with small model size. CDSN is 0.17% better than Arcface, from this point, the discriminative ability of CDSN outperforms its competitors.

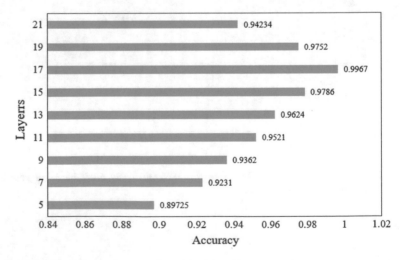

Fig. 5 The number of convolution layers of CDSN and its performance

Table 1 Face verification accuracy of different algorithms on LFW database

Methods	Number of training pairs (M)	Model size (MB)	LFW ACCURACY (%)
Deepface	4	500	97.15
LCNN	4	4	98.92
Facenet	4.7	30	99.08
Mobilefacenet	3.8	4	99.15
Arcface	3.8	112	99.50
CDSN	3	124	99.67

3.5 Metric of Discrimination on the Real-World Database

In order to test the verification accuracy, we test the above algorithms on the real-world database. Here, we randomly select 300 pairs of samples (the same person, positive samples denotes as $(x_i, y_i, 1), i = 1, 2, \ldots, n$) from our real-world face database including 3300 samples. The other 300 pairs of negative sample pairs, write as $(x_j, y_j, 0), j = 1, 2, \ldots, n$). As shown in Fig. 6, two different metric models (euclidean distance and cosine distance) are used to show the accuracy of different methods.

No matter what metric is selected, CDSN yields the best performance on the real-world database.

In order to show the discrimination of features, in this paper, we define a metric to measure the discriminative ability of different algorithms. We directly use a score

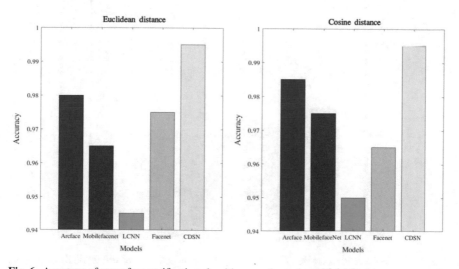

Fig. 6 Accuracy of some face verification algorithms on the real-world database

between 0 and 1 to show the similarity of two samples in semantic space. That is, for same person the score is to be 1, otherwise the score tends to be 0. However, the cosine similarity is in $[-1, 1]$, we should to normalize the similarity scores into $[0, 1]$ by following equation:

$$T_i^{(l)} = \frac{1}{2} + \frac{cos(f(x_i)_k, f(y_i)_k)}{2}, \tag{7}$$

where $T_i^{(l)}$ represents the result of normalization $[0, 1]$, l represents label, which value only with 0 or 1, $f(x_i)_k$, $f(y_i)_k$ represents the image x_i, y_i feature vector extracted by the network, $cos(f(x_i)_k, f(y_i)_k)$ represents the similarity of the extracted features, and the value range is $[-1, 1]$. In order to more clearly describe the ability of each model to distinguish whether two facial images are the same person's, we introduce a metric named as "$score$" as:

$$score = \frac{1}{n}(\sum_{i=1}^{n} T_i^{(1)} - \sum_{i=1}^{n} T_i^{(0)}), (i = 1, 2, \ldots, n). \tag{8}$$

Obviously, the larger the score, the stronger the discriminative ability of the feature which can distinguish whether two face images are the same person. Thus we display the results in Fig. 7.

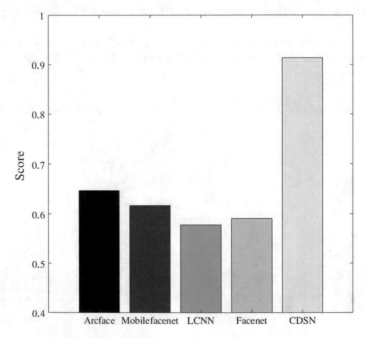

Fig. 7 The discrimination metric of different algorithms on 600 pairs of real world facial images, different colors represent different models, obviously CDSN has the best discriminative scores

From these findings, we confirm the role of the proposed context-aware Siamese neural network.

4 Conclusion

In this paper, we propose a context-aware based discriminative Siamese neural network for face verification. This method incorporates the content-aware model and discriminative Siamese neural network for enhancing the discrimination of learned features. The experimental result shows that the proposed method outperforms other state-of-the-art methods on face verification task.

Acknowledgments This work is supported by the National Natural Science Foundation of China (61502354, 61771353), Central Support Local Projects of China (2018ZYYD059), the Natural Science Foundation of Hubei Province of China (2014CFA130, 2015CFB451), Scientific Research Foundation of Wuhan Institute of Technology (K201713), The 10th Graduate Education Innovation Fund of Wuhan Institute of Technology (CX2018213), 22020 Hubei Province High-value Intellectual Property Cultivation Project, Wuhan City Enterprise Technology Innovation Project (202001602011971), the National Natural Science Foundation of China (62072350).

References

1. Ahonen T, Hadid A, Pietikainen M (2006) Face description with local binary patterns: application to face recognition. IEEE Trans Pattern Anal Mach Intell 12:2037–2041
2. Baker S, Kanade T (2000) Hallucinating faces. In: 2000 the fourth international conference on automatic face and gesture recognition (FG'2000), pp 83–88
3. Chen D, Cao X, Wen F, Sun J (2013) Blessing of dimensionality: high-dimensional feature and its efficient compression for face verification. In: Proceedings of the IEEE conference on computer vision and pattern recognition, pp 3025–3032
4. Chen S, Liu Y, Gao X, Han Z (2018) Mobilefacenets: efficient CNNS for accurate real-time face verification on mobile devices. In: Chinese conference on biometric recognition. Springer, Berlin, pp 428–438
5. Chopra S, Hadsell R, LeCun Y (2005) Learning a similarity metric discriminatively, with application to face verification. In: Null. IEEE, pp 539–546
6. Dalal N, Triggs B (2005) Histograms of oriented gradients for human detection
7. Deng J, Guo J, Xue N, Zafeiriou S (2018) Arcface: additive angular margin loss for deep face recognition (2018). arXiv:1801.07698
8. Guo Y, Lei Z, Hu Y, He X, Gao J (2016) Ms-celeb-1m: a dataset and benchmark for large-scale face recognition (2016)
9. Hu J, Lu J, Tan YP (2014) Discriminative deep metric learning for face verification in the wild. In: Proceedings of the IEEE conference on computer vision and pattern recognition, pp 1875–1882
10. Huang GB, Mattar M, Berg T, Learned-Miller E (2008) Labeled faces in the wild: a database forstudying face recognition in unconstrained environments. In: Workshop on faces in'Real-Life'Images: detection, alignment, and recognition
11. Koch G, Zemel R, Salakhutdinov R (2015) Siamese neural networks for one-shot image recognition. In: ICML deep learning workshop, vol 2

12. Krizhevsky A, Sutskever I, Hinton GE (2012) Imagenet classification with deep convolutional neural networks. In: Advances in neural information processing systems, pp 1097–1105
13. Lee K, Chung Y, Byun H (2002) Svm-based face verification with feature set of small size. Electron Lett 38(15):787–789
14. Lu H, Li Y, Chen M, Kim H, Serikawa S (2018) Brain intelligence: go beyond artificial intelligence. Mob Netw Appl 23(2):368–375
15. Lu H, Li Y, Uemura T, Kim H, Serikawa S (2018) Low illumination underwater light field images reconstruction using deep convolutional neural networks. Futur Gener Comput Syst 82:142–148
16. Lu J, Plataniotis KN, Venetsanopoulos AN, Li SZ (2006) Ensemble-based discriminant learning with boosting for face recognition. IEEE Trans Neural Netw 17(1):166–178
17. Maaten Lvd, Hinton G (2008) Visualizing data using t-SNE. J Mach Learn Res 9(Nov):2579–2605
18. Nair V, Hinton GE (2010) Rectified linear units improve restricted boltzmann machines. In: Proceedings of the 27th international conference on machine learning (ICML-10), pp 807–814
19. Schroff F, Kalenichenko, D, Philbin, J (2015) Facenet: a unified embedding for face recognition and clustering. In: Proceedings of the IEEE conference on computer vision and pattern recognition, pp 815–823
20. Serikawa S, Lu H (2014) Underwater image dehazing using joint trilateral filter. Comput Electr Eng 40(1):41–50
21. Taigman Y, Yang M, Ranzato M, Wolf L (2014) Deepface: closing the gap to human-level performance in face verification. In: Proceedings of the IEEE conference on computer vision and pattern recognition, pp 1701–1708
22. Wang, M., Deng, W.: Deep face recognition: a survey. arXiv preprint arXiv:1804.06655 (2018)
23. Wright J, Yang AY, Ganesh A, Sastry SS, Ma Y (2009) Robust face recognition via sparse representation. IEEE Trans Pattern Anal Mach Intell 31(2):210–227
24. Wu X, He R, Sun Z, Tan T (2018) A light cnn for deep face representation with noisy labels. IEEE Trans Inf Forensics Secur 13(11):2884–2896
25. Zhang K, Zhang Z, Li Z, Yu Q (2016) Joint face detection and alignment using multitask cascaded convolutional networks. IEEE Signal Process Lett 23(10):1499–1503

Object-Level Matching for Multi-source Image Using Improved Dictionary Learning Algorithm

Xiong Zhenyu, Lv Yafei, Zhang Xiaohan, Zhu Hongfeng, Gu Xiangqi, and Xiong Wei

Abstract Solving the problem of multi-source information matching is the foundation of multi-source information fusion. Aiming at the heterogeneity between different multi-source images, we propose a new method of object-level matching for multi-source image based on improved dictionary learning. Two main steps, unified representation and similarity measure, are contained. Firstly, we complete the unified representation of multi-source images by improved dictionary learning algorithm. On the basis of the representation ability of dictionary learning, we further make full use of the label information to boost the discriminative ability of dictionary, which is beneficial to the implementation of object matching. Then, we construct a neural network to learn the distance metric standard between matching and non-matching by supervised learning, which can replace the traditional distance metric method. In addition, we produce two sets of multi-source image object matching datasets based on the open datasets, which verifies the validity and accuracy of the algorithm, and shows the good performance of the algorithm in solving the zero-shot learning problems.

Keywords Dictionary learning · Object-level matching · Multi-source images · K-SVD algorithm · Zero-shot learning

1 Introduction

With the increasing of multi-source sensors and multi-source data, we have entered an era of remote sensing big data and new methods of processing and understanding the information behind the big data are needed urgently. Object-level matching in multi-source images, one of the fundamental techniques for mining the big data, is the basis of fusing and utilizing multi-source information. In practical applications, the indirect matching is mainly used. Firstly, the object feature information is used to judge the object's category, and then the matching is conducted according to whether

X. Zhenyu (✉) · L. Yafei · Z. Xiaohan · Z. Hongfeng · G. Xiangqi · X. Wei
Research Institute of Information Fusion, Naval Aviation University, Yantai, China
e-mail: 137908892@qq.com

© Springer Nature Switzerland AG 2021
H. Lu (ed.), *Artificial Intelligence and Robotics*,
Studies in Computational Intelligence 917,
https://doi.org/10.1007/978-3-030-56178-9_10

the judged object's category is consistent. However, the main disadvantages of this method are as follows: the accuracy of matching is overwhelmingly dependent on the completeness of the identification databases and the accuracy of the algorithm; the object feature is first extracted for judging the category of object and then the matching result is acquired by the category information, which increases the loss of information. Besides, the most studied multi-source image fusion is to achieve image fusion in the pixel-level to obtain more detailed information, which is convenient for further analysis, processing and understanding of images, but it has high requirements for temporal aligning and spatial aligning of input images, and has a large amount of calculations and poor real-time performance. Considering the diversity of multi-sensor platform and the heterogeneity of images, we intend to analyze the feature information of multi-source images by extracting and comparing the feature information to determine whether the images originate from the same object and construct a mapping relationship between objects in multi-source images.

However, there are few corresponding research due to the high dimensionality of feature information and the large difference in the feature information of heterogeneous sources. In order to overcome this limitation, our motivations are as follows and shown in Fig. 1: feature extraction and similarity measure are usually adopted to conduct object matching. The objects in images can be mapped into feature representations in a common feature space, and the similarity can be measured by calculating the distance between the feature representations. But for multi-source images, the existing data drift problem prevents the implementation of similarity measure. So, how to perform a unified representation of the multi-source images to eliminate the data drift problem is crucial.

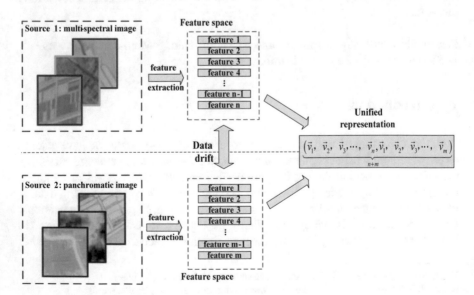

Fig. 1 The motivation of the paper

In this paper, we skip the object recognition and directly use the feature representations between multi-source images to make matching. As discussed above, we introduce the dictionary learning method into object-level matching, which is used to unify the multi-source image in the same description space. The problem of the object-level matching for multi-source image can be divided into two stages: the representation learning phase and the similarity measure phase. In the stage of representation learning, a unified dictionary of multi-source images is generated by proposing an improved dictionary learning algorithm, which utilizes further the label information in the objective function while considering the representation ability of the dictionary, so that the sparse coefficients of the matching object are as close as possible, and the sparse coefficients of the non-matching object are as different as possible. Furthermore, a distance metric can be learned based on the sparse coefficients generated by the unified dictionary. Specially, a matching discriminative neural network is constructed to learn the relation between the sparse coefficients and label information in a supervised way. Finally, we use this distance metric to judge whether the object information acquired by the two sources comes from the same object, and realize multi-source image object matching. The framework of our improved dictionary learning algorithm is shown in Fig. 2.

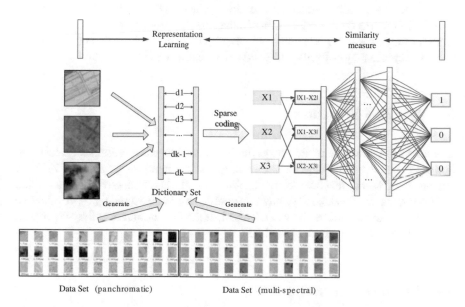

Fig. 2 Framework of our improved dictionary learning algorithm

2 Related Work

Dictionary learning (sparse representation) has been successfully applied in various fields such as computer vision and signal decomposition, especially in image classification [1, 2], image denoising [3], image recognition [4, 5]. As a classic dictionary generation method, K-SVD [6] algorithm has become the preferred method for generating dictionary in dictionary learning since its introduction, with its excellent representation ability and good convergence speed. However, in the process of image classification, due to the insufficient classification ability of the dictionary, the classification effect is suppressed. On this basis, many supervised learning algorithms are proposed to improve the classification ability of the K-SVD algorithm dictionary. On the basis of K-SVD, the literature [4] introduces the classification error into the objective function, and proposes the D-KSVD algorithm. In the process of optimization, the dictionary's representation ability and classification ability are optimized at the same time, and good results are obtained in face recognition. Further, Literature [5] introduces label information into the objective function by considering the classification error, and proposes the LC-KSVD algorithm. So that the sparse coefficients of similar labels tend to be similar, the sparse coefficients of different types of labels tend to be different, and it is easier to classify; In [7], the label pair error is introduced into the dictionary learning framework, considering reconstruction error, classification error and label pair error in the objective function, so that the dictionary has both representation ability and good classification ability.

3 Algorithm Structure

The algorithm structure of this paper includes two stages: representation learning and discriminative matching. As shown in Fig. 2, the input image data Y_1, Y_2, Y_3 is characterized as a sparse coefficient X_1, X_2, X_3 in the same space through a unified dictionary, and the "distance" of any two coefficients is measured by discriminative model to output the number between 0 and 1. If the final output is 1, the two objects are matched; if 0, they are not matched.

3.1 Dictionary Learning

Dictionary learning is to find a complete dictionary $D = [d_1, \ldots, d_K] \in R^{n \times K} (K > n)$ for sample $Y = [y_1, \ldots, y_n] \in R^{n \times N}$, and to represent the sample as a linear combination of the elements in the dictionary, to ensure that the coefficients of the linear combination $X = [x_1, \ldots, x_N] \in R^{K \times N}$ are as sparse as possible, and to control the extent of information loss, thereby simplifying learning tasks and reducing the complexity of the model. It can be seen from the above definition that dictionary learning should achieve the following learning objectives: learning to obtain a complete dictionary, the sparse representation of the coefficients should be as sparse as possible, and the reconstruction error should be controlled within a certain range. The mathematical expression is shown in formula (1):

$$< D, X >= \arg\min_{D,X} \sum_{i=1}^{N} (||y_i - Dx_i||_2^2 + \alpha||x_i||_1) \tag{1}$$

Among them, the first item in the formula is the linear combination of the control dictionary and the sparse coefficient to restore the sample set as much as possible to control the reconstruction error; the second item uses the L1- norm to control the sparseness of the sparse coefficient; for the hyper-parameter α, it can be used to balance the weight of reconstruction error and the degree of sparsity.

It can be known from formula (1) that dictionary learning should generate dictionary D and sparse coefficient X through unsupervised learning. The generation of dictionary is the basis of the algorithm. And the performance of dictionary determines the performance of dictionary learning algorithm. The optimization of formula (1) is a non-convex optimization problem for D and X, but the problem of fixing one parameter and optimizing another parameter is the convex optimization problem. Therefore, dictionary learning is generally divided into two-step solving. First, the dictionary is fixed, the sparse coefficients are solved, then the sparse coefficients are fixed, the dictionary is updated, and the iteration is repeated to the optimal. The difference between the dictionary learning algorithms lies in the difference between the method of solving the sparse coefficients and the method of updating the dictionary.

Solving the sparse coefficient in the fixed dictionary can be proved as NP-hard problem [8], which can be solved by Orthogonal Matching Pursuit (OMP) [9] and Basis Pursuit (BP) [10]. For a dictionary generated by fixed sparse coefficients, the commonly used methods are the Method Of Optimal Directions (MOD) [11], and FOCUSS method [12].

3.2 K-SVD Algorithm

The K-SVD algorithm can be regarded as a generalization of the K-means algorithm. The K-means algorithm is to find an approximate representation for each signal element, while the K-SVD algorithm is to approximate each signal with a linear combination of multiple elements.

When the K-SVD algorithm generates a dictionary, each iteration only updates one dictionary element, looking for an optimal $\tilde{d}_k$ and updating the corresponding sparse coefficients $\tilde{x}_T^k$ ($\tilde{x}_T^k$ is the kth row of the coefficient matrix X, corresponding to the kth row of the dictionary d_k) to minimize the formula (2). Therefore, the difference from the other algorithms for generating dictionary is that the K-SVD algorithm changes the sparse coefficients while updating the dictionary, and after all the dictionary elements are updated, The K-SVD algorithm uses any dictionary to update the global coefficients to avoid falling into local optimum.

$$||Y - DX||_F^2 = \left\| Y - \sum_{j=1}^{K} d_j x_T^j \right\|_F^2$$

$$= \left\| \left(Y - \sum_{j=1}^{K} d_j x_T^j \right) - d_k x_T^k \right\|_F^2$$

$$= \left\| E_k - d_k x_T^k \right\|_F^2 \qquad (2)$$

If you directly perform a singular value decomposition (SVD) on E_k to update d_k and x_T^k, there may be a phenomenon that the sparse coefficient x_T^k is not sparse. For this reason, the literature [6] only retain non-zero values in the sparse coefficient x_T^k to form E_R^k, and performs singular value decomposition on E_R^k to obtain: $E_R^k = U \Lambda V^T$.

3.3 Characterization Learning-Generation of Multi-source Dictionary

In order to make the sparse coefficient have certain discriminative ability, the sparse coefficient of the matching object is as similar as possible, while the sparse coefficient of the non-matching object is as different as possible. Based on the objective function of dictionary learning, we add a label information to get the objective function as shown in formula (3),

$$< D, X > = \arg\min_{D,X} \sum_{i=1}^{N} (||y_i - Dx_i||_2^2 + \alpha||x_i||_1) + \frac{\beta}{2} \sum_{i,j}^{N} (||x_i - x_j||_2^2 M_{ij})$$

$$= \arg\min_{D,X} \sum_{i=1}^{N} (||y_i - Dx_i||_2^2 + \alpha||x_i||_1) + \beta(Tr(X^T XD) - Tr(X^T XM))$$

$$= \arg\min_{D,X} \sum_{i=1}^{N} (||y_i - Dx_i||_2^2 + \alpha||x_i||_1) + \beta(Tr(X^T XL)) \qquad (3)$$

where α and β are hyper-parameters, which are used to control the weight of the error represented by each item; the third term is a restriction item that makes the sparse coefficient have the classification ability. As shown in formula (4), when the label of the object pair (y_i, y_j) is matching object, $M_{ij} = 1$; when the label (y_i, y_j) of the object pair is non-matching object, $M_{ij} = -1$; other cases, such as the same object (y_i, y_j), $M_{ij} = 0$. $D = diag\{d_1, \ldots, d_N\}$ is a diagonal matrix, each diagonal

element is the sum of the elements of each column in the matrix M, $d_i = \sum_{j=1}^{N} M_{ij}$;
$L = D - M$.

$$M_{ij} = \begin{cases} +1, \ if \ (y_i, y_j) \in S \\ -1, \ if \ (y_i, y_j) \in D \\ 0, \ otherwise \end{cases} \tag{4}$$

3.4 Optimization Method

Like the formula (1), the objective function formula (3) is a non-convex optimization problem for the parameters D and X. It is necessary to fix one parameter to solve another parameter, and thus the optimization problem of the objective function can be converted into iterative optimization of the following two formulas:

$$L(x_i) = \arg\min_{x_i} ||y_i - Dx_i||_2^2 + \alpha||x_i||_1 + \beta(2x_i^T X L_i - x_i^T x_i L_{ii}) \tag{5}$$

$$L(D) = \arg\min_{D} \sum_{i=1}^{N} ||y_i - Dx_i||_2^2 = \arg\min_{D} ||Y - DX||_F^2 \tag{6}$$

Among them, the formula (5) optimizes the sparse coefficient X for the fixed dictionary D, and the formula (6) optimizes the dictionary set for the fixed sparse coefficient. According to the formula (2), Eq. (6) can be optimized with K-SVD, by updating the elements in the dictionary and the corresponding sparse coefficients one by one; for the formula (5), the objective function is not continuously differentiable due to the existence of the L1-norm, but the method for solving the problem is given in [13]. The "feather-sign search algorithm" method proposed in the literature transforms the object function into a standard, unconstrained quadratic optimization problem (QP) by iteratively generating the sign vector θ of the corresponding sparse coefficient. Therefore, the gradient of formula (5) is calculated, as shown in formula (7),

$$\frac{\partial L(x_i)}{\partial x_i} = 2D^T (Dx_i - y_i) + 2\beta X L_i + \alpha\theta \tag{7}$$

Let the result of formula (7) equals to zero, and the sparse coefficient can be obtained, as shown in the formula (8).

$$x_i^* = (D^T D + \beta L_{ii} I)^{-1} \cdot (D^T y_i - \frac{1}{2}\alpha\theta) \tag{8}$$

3.5 Discriminative Model

After generating a dictionary to sparsely represent multi-source image information, model each sparse coefficient, and use the tag information match with the object to establish an discriminative model, and convert the matching problem into a binary classification problem. The discriminative model uses a multi-layer neural network model in this paper.

The sparse coefficients and labels of each object are preprocessed to form a ternary dataset (x_i, x_j, y_{ij}). x_i and x'_j represent sparse representations of the two sources and y_{ij} is the result of manual marking. $y_{ij} = 0$ indicates that x_i and x'_j are non-matching objects, and $y_{ij} = 1$ indicates that x_i and x'_j are matching objects. Because the matching problem can be regarded as a binary classification, the sigmoid function is used in the output layer. After multiple trials, the network structure is set to a four-layer network. The activation function of the middle layer uses the relu function to add dropout to each layer of the network to prevent overfitting.

4 Experiments and Analysis

The experimental part of this paper selects public remote sensing image datasets, including two types of image data: multi-spectral image and panchromatic image. In this paper, two types of images are preprocessed, and the same region or the same object is respectively cut to form multi-source image matching datasets. A partial example of multi-source image matching datasets is shown in Fig. 3.

4.1 The Ability of Dictionary to Characterize Multi-source Images

We made multi-source image matching datasets based on the remote sensing image datasets, in which the full-color image and the multi-spectral image respectively

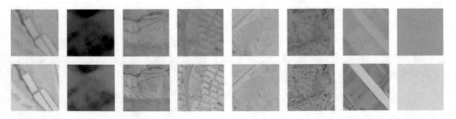

Fig. 3 Multi-source remote sensing image dataset (the first line is a multi-spectral image and the second line is a panchromatic image.)

contain 1824 images, and 1600 of them are selected as training sets for generating common dictionary and matching discriminative model, the remaining 224 images were used as test sets.

Figures 4 and 5 show the reconstruction of panchromatic and multi-spectral images in the training set and test set. It can be seen from Fig. 4 that the reconstruction of the training set of the dictionary is very close to the original image and difficult to distinguish with the naked eye; in Fig. 5, the reconstruction result of the dictionary has a certain gap with the original image, and the contour lines are basically restored. However, the details are generally lost and blurred; from the following images, the dictionary learning and the proposed method are similar in performance to the training set and test set reconstruction, indicating that the two dictionary are similar in terms of representation ability, which is consistent with the starting point of this paper, maintaining the representation ability of the dictionary and increasing the certain discriminative ability.

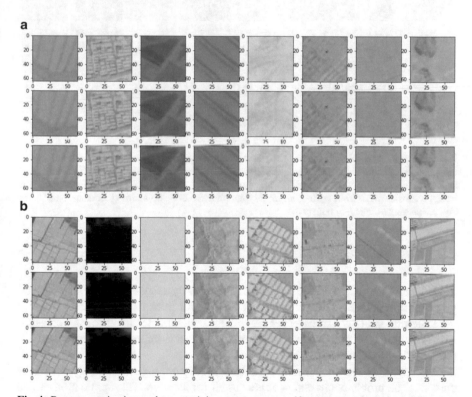

Fig. 4 Remote sensing image dataset training set reconstructed image comparison chart. **a** Comparison of multi-spectral image reconstruction (from top to bottom: original image, dictionary learning reconstructed image and reconstructed image of our method). **b** Contrast map of panchromatic image reconstruction (from top to bottom: original image, dictionary learning reconstructed image and reconstructed image of our method)

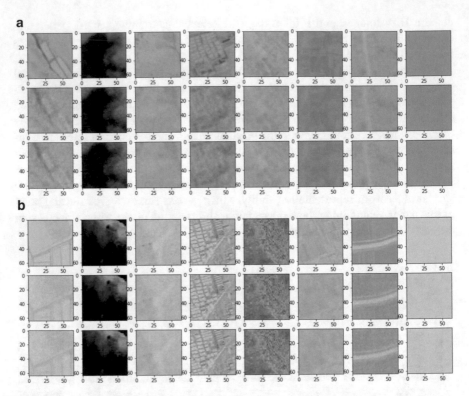

Fig. 5 Test set of remote sensing image dataset reconstructed image comparison chart. **a** Comparison of multi-spectral image reconstruction (from top to bottom: original image, dictionary learning reconstructed image and reconstructed image of our method). **b** Contrast map of panchromatic image reconstruction (from top to bottom: original image, dictionary learning reconstructed image and reconstructed image of our method)

4.2 The Accuracy of Matching

Table 1 show the accuracy of labeled matching pairs by using dictionary learning and proposed approach in multi-source image dataset. It can be seen from the results that the matching strategy of this paper has a high matching accuracy for the datasets, and the accuracy of proposed approach on the test set is improved by more than 8% compared with the dictionary learning.

While we have thus far considered multi-source image object matching as a binary classification problem, our end goal is to use it for location. This application can be

Table 1 Accuracy of labelled matching pairs in multi-source image dataset (%) (p = 224)

Method	Accuracy (%)
Dictionary learning	84.17
Proposed approach	92.86

Table 2 The multi-spectral image dataset matching scores (%) (p = 224)

Method	Rank = 1	Rank = 5	Rank = 10	Rank = 20
Dictionary learning	24.32	54.98	65.95	83.67
Proposed approach	29.67	58.56	69.04	87.53

framed as a ranking or retrieval problem: given a query image and a repository of images, one of which is the match, we want to rank the images according to their relevance to the query so that the true match image is ranked as high as possible. The ranking task is typically approached as following: the representation learning networks are applied to the query image and the repository's images to obtain their feature vectors. Then these images can be ranked by sorting the distance from their features to the query image's feature. The localization is considered successful if the true match image is ranked within a certain top percentile.

Table 2 show the accuracy of the cumulative matching scores (%) between the dictionary learning and the proposed approach in multi-source image dataset at rank 1, 5, 10, and 20. It can be seen from the results that the accuracy of proposed approach on the test set is improved by more than 5% compared with the dictionary learning.

4.3 Generalization of Dictionary-Zero Sample Learning Problems

In practical applications, considering the heterogeneity of multi-source image information, it is difficult to ensure that there are sufficient samples of multi-source image data when generating a unified dictionary. Whether the dictionary has good representation ability in the face of the "unseen" image set (the problem of zero sample learning), and thus the corresponding matching ability needs to be tested. The DSTL satellite image dataset is a publicly available multi-source image dataset that includes two types of satellite image spectral data: a 3-band RGB visible image and a 16-band hyper-spectral image. In this paper, we use the dictionary of multi-source remote sensing image dataset to reconstruct the samples of DSTL dataset. The results are shown in Fig. 6. It can be seen from the figure that although the reconstruction effect is fuzzier, the image contour is basically reconstructed. It has a certain identifiability, which verifies the good generalization ability and zero sample learning ability of the proposed method in multi-source image object matching.

Tables 1 show the accuracy of labeled matching pairs by using dictionary learning and proposed approach in multi-source image dataset. It can be seen from the results that the matching strategy of this paper has a high matching accuracy for the datasets, and the accuracy of proposed approach on the test set is improved by more than 5% compared with the dictionary learning.

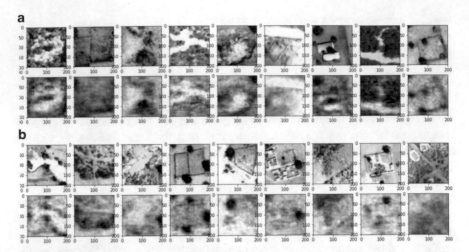

Fig. 6 Multi-source remote sensing image dictionary reconstruction DSTL dataset comparison chart. **a** Comparison of multi-spectral image reconstruction (from top to bottom: original image, reconstructed image). **b** Comparison of visible light image reconstruction (from top to bottom: original image, reconstructed image)

Table 3 The multi-spectral image dataset matching scores (%) (p = 224)

Method	Accuracy (%)
Dictionary learning	78.17
Proposed approach	83.86

5 Summary

In this paper, we propose a new method for multi-source image object matching, which unifies the representation of multi-source image by improving the dictionary learning algorithm. We use the neural network to learn the distance metric between the objects, and then use this standard on the two sets of self-made datasets. The validity of the algorithm and the high accuracy of matching are tested. Finally, the method is proved to have good applicability in zero-shot learning.

Though our proposed object matching approach can achieve better performance, there are still some shortcomings that we cannot neglect. When we face more different types of images, the training time of the model is long and the accuracy of the object matching will be influenced. For zero-shot learning, it's hard to distinguish the detail of the object. So, how to overcome the limitation for the use of module is one of our future focuses.

References

1. Wang J, Yang J, Yu K, et al (2010) Locality-constrained linear coding for image classification. Computer vision and pattern recognition. IEEE, 2010:3360–3367
2. Vu TH, Mousavi HS, Monga V et al (2016) Histopathological image classification using discriminative feature-oriented dictionary learning. IEEE Trans Med Imaging 35(3):738–751
3. Elad M, Aharon M (2006) Image denoising via sparse and redundant representations over learned dictionaries. IEEE Trans Image Process15(12):3736–3745
4. Zhang Q, Li B (2010) Discriminative K-SVD for dictionary learning in face recognition. Computer vision and pattern recognition. IEEE, 2010:2691–2698
5. Jiang Z, Lin Z, Davis LS (2013) Label consistent K-SVD: learning a discriminative dictionary for recognition. IEEE Trans Softw Eng 35(11):2651–2664
6. Aharon M, Elad M, Bruckstein A (2006) K-SVD: an algorithm for designing overcomplete dictionaries for sparse representation. IEEE Trans Signal Process 54(11):4311–4322
7. Guo H, Jiang Z, Davis LS (2012) Discriminative dictionary learning with pairwise constraints. In: Asian conference on computer vision. Springer, Berlin, Heidelberg, 2012:328–342
8. Davis G, Mallat S, Avellaneda M (1997) Adaptive greedy approximations. J Construct Approx 13:57–98
9. Pati YC, Rezaiifar R, Krishnaprasad PS (1993) Orthogonal matching pursuit: recursive function approximation with applications to wavelet decomposition 1:1–3
10. Chen SS, Donoho DL, Saunders MA (2000) Atomicdecomposition by basis pursuit. SIAM Rev 43(1):129–159
11. Engan K, Aase SO, Husøy JH (2000) Multi-frame compression: theory and design. EURASIP Signal Process 80(10):2121–2140
12. Gorodnitsky IF, Rao BD (1997) Sparse signal reconstruction from limited data using FOCUSS: a re-weighted norm minimization algorithm. IEEE Trans Signal Process 45:600–616
13. Lee H, Battle A, Raina R, Ng AY (2007) Efficient sparse coding algorithms. Neural Inf Process Syst

Classification of Hyperspectral Image Based on Shadow Enhancement by Dynamic Stochastic Resonance

Xuefeng Liu, Hao Wang, Min Fu, and Bing Zheng

Abstract Information extraction of shadow areas in hyperspectral images (HSIs) has always been a difficult problem in HSI processing. Dynamic stochastic resonance (DSR) theory has proved that the noise contained in the signal can enhance the strength of the original signal and improve the signal-to-noise ratio (SNR). And it has been applied in signal and image processing,communication and other fields. In this paper, DSR theory is introduced to the shadow enhancement in HSIs for the first time. The spatial and spectral dimensions of the shadow areas in a HSI could be enhanced by the DSR respectively. Then, the enhanced shadow should be fused with the other areas in the HSI. Finally, the fused image could be classified to explore the information in the HSI. The experimental result show that the DSR has promising prospect in the shadow enhancement in HSIs, and can help to improve the classification.

Keywords Remote sensing images · Stochastic resonance · Support vector machine · Shadow area · Parameter adjustment

1 Introduction

Hyperspectral images (HSIs) are acquired by hyperspectral sensors or imaging spectrometers, which use tens to hundreds or even more continuous bands to simultaneously image the target space, fuse the image with the spectrum, and obtain spatial and spectral information at the same time [1]. With the further development of hyperspectral imaging system, the spatial and spectral information contained in HSI will be more abundant and accurate. Because of the characteristics of HSI, hyperspectral

X. Liu · H. Wang
Qingdao University of Science and Technology, No. 99 Songling Road, Qingdao, China
e-mail: nina.xf.liu@hotmail.com

M. Fu (✉) · B. Zheng
Ocean University of China, No. 238 Songling Road, Qingdao, China
e-mail: fumin@ouc.edu.com

© Springer Nature Switzerland AG 2021
H. Lu (ed.), *Artificial Intelligence and Robotics*,
Studies in Computational Intelligence 917,
https://doi.org/10.1007/978-3-030-56178-9_11

remote sensing earth observation technology has been widely used in many differ-ent fields, such as mining, astronomy, chemical imaging, agriculture, environmental science, wildland fire tracking and biological threat detection and so on [2].

The main purpose of HSI classification is to classify the pixels in the image so as to realize the automatic classification of objects [3]. However, due to the limitation of weather and terrain environment in which hyperspectral imaging instruments work, the information contained in HSI will be polluted by noise or completely submerged. Especially in the shadow area of HSI, due to insufficient illumination, refraction and scattering of light, the effect of detection, recognition and classification of objects in HSI is reduced, The shadow in HSIs is one of the main difficulties in information mining. Dynamic Stochastic Resonance (DSR) has been proved to be a good enhancement for shaded areas in gray and color images. Based on this, a DSR technique is proposed to enhance the shaded area of HSI in order to facilitate the detection and recognition of objects in the shaded area of HSI.

Stochastic Resonance (SR) theory was put forward by Benzi et al. in 1981 [4] when he explained the phenomena of glacial and warm climatic periods alternately in paleo meteorology. SR enhanced the signal by the interaction of non-linear system, weak driving signal and noise. Although the development of SR theory is not long, scholars have confirmed the existence of SR phenomenon in meteorological, biological and circuit systems etc. [5]. SR theory has been applied to cochlear implant design, detector enhancement [6], image enhancement [7] and signal processing as a typical theory to effectively utilize noise energy [8]. It has even been applied to unexpected areas, such as mechanical fault detection [9] and the human vestibular system [10].

The remainder of this paper is organized as follows: Sect. 2 overviews the main theory of DSR; the proposed method is introduced in Sect. 3 in detail; Sect. 4 presents the experimental results and discussion; the conclusion is contained in Sect. 5.

2   Overview of DSR

In signal processing or other related research work, the noise of interference signal usually troubles researchers. It will distort the signal and greatly reduce the validity and correctness of the signal. However, recent studies on stochastic resonance show that the noise existing in weak signal is entirely possible to amplify weak signal. In other words, the noise in weak signal can play an important role in enhancing signal and improving signal-to-noise ratio.

The classical stochastic resonance theory can be described by the motion of an overdamped particle between bistable potential wells [11]. Suppose that a particle with mass m and friction coefficient γ moves in a bistable potential well determined by $U(x)$ under over-damped condition and the particle is affected by noise $\xi(t)$ and periodic driving signal $f(x)$, then it's Langevin equation of motion is:

$$m\frac{d^2x(t)}{dt} + \gamma\frac{dx(t)}{dt} = -\frac{dU(x)}{dx} + f(t) + \xi(t) \tag{1}$$

Fig. 1 Bistable double-well potential system ($a = 1$, $b = 1$)

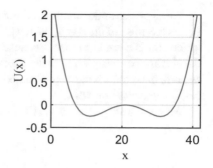

where $x(t)$ is the position of the particle, $\frac{dx(t)}{dt}$ is the rate of change of the particle position. If the model is seriously over-damped, the equation can be simplified as:

$$\frac{dx(t)}{dt} = -\frac{dU(x)}{dx} + f(t) + \xi(t) \tag{2}$$

where $U(x)$ is a bistable fourth-order potential well function with mirror symmetry as shown in Fig. 1:

$$U(x) = -a\frac{x^2}{2} + b\frac{x^4}{4} \tag{3}$$

with a and b being the parameters of the bistable system. The potential well has two stable states located at $x_\pm = \pm\sqrt{\frac{a}{b}}$ and one unstable state at $x_s = 0$. Between the stable state and the unstable state, there is a potential barrier with a height of $\Delta U = \frac{a^2}{4b}$.

Based on the theory above, Langevin equation describes the movement state of particles moving back and forth between two potential wells under the combined action of periodic driving signal and noise. When there is no periodic signal driving, the particles oscillate around the stable state with small amplitude, and the statistical variance of the oscillation amplitude is proportional to the noise intensity. Under the action of periodic driving signals, the symmetrical bistable potential well sways asymmetrically with the amplitude of periodic signals, i.e. the uneven potential wells on both sides, the particles begin to oscillate greatly, but the driving signal amplitude is relatively small, because the particles can not cross the barrier, only the periodic signal is not enough to make the particles jump periodically in the two potential wells. Adding appropriate intensity of noise can help the particles jump from one potential well to another according to the frequency of the periodic signal. The original chaotic noise acts on the bistable system and produces some statistical characteristics in the form of the motion state of the output particles.

In order to apply stochastic resonance theory to HSI processing, based on the above model, we further deduce Eq. (3) into Eq. (2):

$$\frac{dx(t)}{dt} = [ax - bx^3] + f(t) + \xi(t) \tag{4}$$

Based on the understanding of stochastic resonance, $f(t)$ is the input signal, and $\xi(t)$ is the noise of the signal. When processing the HSI, we can think that the shadow in the HSI is a kind of noise. Because of the influence of the noise, the image under the shadow becomes very weak. Therefore, the three conditions of DSR (a. bistable system. b. weak driving signal. c. external noise.) have been satisfied, and the HSI can be processed by DSR.

We use the shadow area in HSI as the input of the DSR system:

$$Input = f(t) + \xi(t) \tag{5}$$

then:

$$\frac{dx(t)}{dt} = [ax - bx^3] + Input \tag{6}$$

In order to make the system suitable for image processing, we rewrite the above differential form to differential form:

$$x(n+1) = x(n) + \triangle t[(ax(n) - bx^3(n)) + Input] \tag{7}$$

where n is the iteration number of dynamic stochastic resonance. The initial value of $x(0) = 0$.

3   Improved HSI Classification Based on Shadow Enhancement by DSR

As mentioned above in Sect. 2, DSR can enhance the signal intensity and signal-to-noise ratio. In HSIs, the shadow area could be taken as a weak signal contaminated by noise. Therefore, DSR could help to enhance the shadow area in HSIs.

In this paper, we propose to enhance the shadow area in HSIs by DSR in the spatial and the spectral dimensions respectively, and then the fused image could be classified by the support vector machine (SVM). The flow chart of the proposed method is shown in Fig. 2.

Firstly, due to only the shadow areas in HSIs needing to be enhanced, we use the shadow extraction mask to extract the shadow areas of the original hyperspectral data and obtain the three-dimension (3D) tensor form of shadow data.

Secondly, DSR is performed in the spatial and the spectral dimensions respectively to enhance the shadow areas.

- DSR on the spatial dimension: the extracted shadow data are divided into several 2D images according to the band. In order to calculate more conveniently by Eq. (7), we need to transform the 2D shadow area images into 1D row vectors, After DSR in Eq. (7) the enhanced 1D vector should be reshape to 2D matrix. Shadow data in the form of 3D tensor after processing are obtained by sequential arrangement.

Fig. 2 Flow chart of the
proposed method

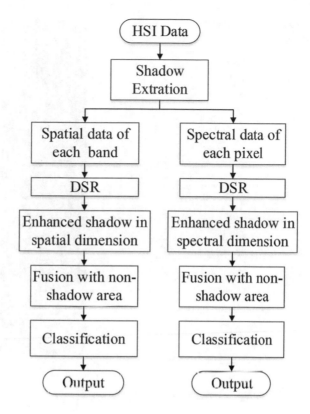

- DSR on the spectral dimension: in the shadow data, the 1D spectral information
  of each pixel can be extracted, and enhanced by the Eq. (7). After all the spectral
  data of all the pixels in shadow areas are processed, the shadow data in the form
  of 3D tensors could obtained.

  Thirdly, the enhanced shadow areas should be fused with the non-shadow areas
  in HSIs to obtain the enhanced image.

  Finally, SVM could be used to classify the enhanced HSI.

4 Experiment

In this paper, the HYDICE (Hyperspectral Digital Imagery Collection Experiment)
HSI which contains a lot of shadows is used in the experiment. As shown in Fig. 3,
this HYDICE HSI has a 0.75 m spatial and 10 nm spectral resolution and includes
148 spectral bands (from 435 to 2326 nm), 316 rows and 216 columns. The image
contains the following objects: grass, road, tree, shadow (including a section of road),
target 1, target 2 and target 3.

Fig. 3 HYDICE HSI

4.1 DSR Parameter Adjustment

It can be seen from the Eq. (7) that DSR mainly contains three parameters a, b, Δt, n. Among them, a and b are parameters in bistable system. They mainly affect the position of potential well and the height of potential barrier in bistable system. Δt is the iteration step in DSR processing, n is the iteration number of DSR processing, Δt and n mainly affect the gain of DSR processing. Because there is no mature theoretical support for the selection of parameters in the current research on dynamic stochastic resonance, we try to find the optimal value of parameters for DSR processing in this paper from the perspective of spectral dimension processing.

To determine the optimal values of a and b, we fixed the value of Δt to 0.01 and n to 24 firstly, and selected a spectral data of road pixels in the shadow from HYDICE HSI to do DSR processing. Then different a and b are selected to use to get different enhanced spectral data. The Pearson correlation coefficients of these data are calculated with the standard road spectra in HYDICE HSI. Pearson correlation coefficient is defined as Eq. (8):

$$r = \frac{\sum_{i=1}^{n}(X_i - \overline{X})(Y_i - \overline{Y})}{\sqrt{\sum_{i=1}^{n}(X_i - \overline{X})^2}\sqrt{\sum_{i=1}^{n}(Y_i - \overline{Y})^2}} \tag{8}$$

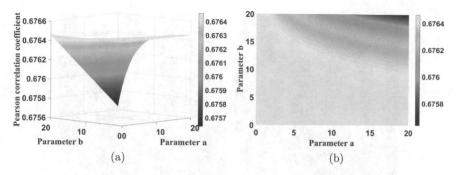

Fig. 4 The influence of the values of a and b on Pearson correlation coefficient. **a** 3D view. **b** 2D view

where X and Y are two 1D vectors with length n, and $\overline{X}\,\overline{Y}$ represent the average of X and Y. Equation (8) shows that the closer the Pearson correlation coefficient is to 1, the higher the correlation between the two data is. On the contrary, the closer the Pearson correlation coefficient is to 0, the worse the correlation between the two data is. The influence of the values of a and b on Pearson correlation coefficient is shown in Fig. 4.

4.2 DSR Shadow Enhancement

According to Sect. 3, the shadows of the HYDICE HSI can be enhanced in the spatial and the spectral dimensions respectively and some of the results are illustrated in Fig. 5 and Fig. 6.

It can be seen from Figs. 5 and 6, the DSR can enhance both the spatial and the spectral data in the shadow area. The spatial dimension DSR processing can enhance the contrast of the shadow area and make some details more prominent under the shadow. The spectral dimension DSR processing can greatly improve the original weak spectral data and make the spectral characteristics of the pixels under the shadow clearer. From these two aspects, DSR can enhance the shadow area of HSI significantly.

4.3 Classification of SVM

In recent years, machine learning and artificial intelligence [12] have developed continuously, and have been applied to image processing [13, 14], recognition [15], anomaly detection [16] and other fields. SVM is a classical method based on supervised learning. In the experiment, we have used SVM to classify the enhanced HSI. In order to compare and verify the enhancement effect of DSR, we chose two region

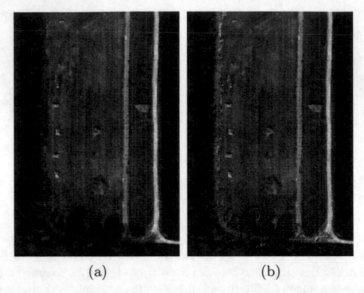

(a)   (b)

Fig. 5 1st band of HYDICE HSI before and after DSR enhancement. **a** Before DSR enhancement. **b** After DSR enhancement

Fig. 6 Comparison of the spectral dimension before and after DSR processing

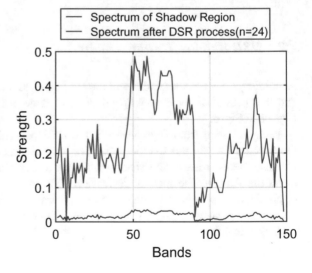

of interest (ROI) selection methods in SVM classification. One contains eight samples (ROI1): road, road under shadow, grass, tree, shadow, target 1, target 2, target 3; the other contains seven samples (ROI2): road and road under shadow, grass, tree, shadow, target 1, target 2 and target 3. Two kinds of ROIs have been used to classify the original HSI, the space-dimensional DSR-enhanced HSI and the spectral-dimensional DSR-enhanced HSI.

To evaluate the improvement of the classification, the overall accuracy (OA) in percentage is applied and defined as: for P class C_i ($i = 1, ..., P$), if a_{ij} is the number of test samples that actually belong to class C_i and is classified into C_j ($j = 1, ..., P$), then

$$OA = \frac{1}{M} \sum_{i=1}^{P} a_{ii} \tag{9}$$

where M is the total number of samples, P is the number of classes C_i and $a_{ii} = a_{ij}$ for $i = j$.

The classification results are shown in Figs. 7 and 8. The accuracy comparison of classification is listed in Table 1.

Because SVM only uses the spatial information of HSI in classification, it is not obvious that the classification accuracy of HSI after processing is improved. From the above results, it can be seen that both spatial dimension and spectral dimension DSR have certain effects on the enhancement of HSI shadow area and the improvement of HSI classification accuracy.

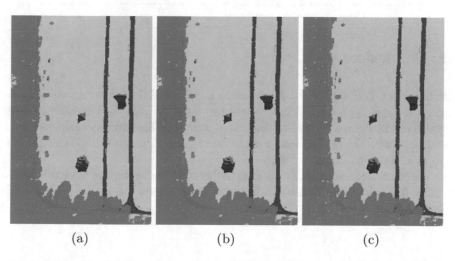

| (a) | (b) | (c) |

Fig. 7 Classification based on ROI1. **a** Original HSI. **b** DSR on spatial dimension. **c** DSR on spectral dimension

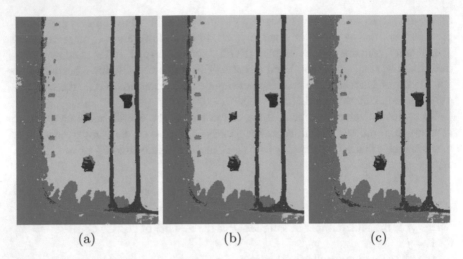

(a)   (b)   (c)

Fig. 8 Classification based on ROI2. **a** Original HSI. **b** DSR on spatial dimension. **c** DSR on spectral dimension

Table 1 Overall accuracy of classification

	Original HSI (%)	DSR on spatial dimension (%)	DSR on spectral dimension (%)
ROI1	90.0727	90.7964	90.5898
ROI2	89.7035	90.4971	90.5957

5 Conclusion

In this paper, DSR theory is firstly introduced into the HSI shadow region enhancement, and it is applied to process the shadow region of HSI from two aspects: the spatial dimension and the spectral dimension. In the classification after enhancement, compared with the original image classification, the proposed method can help to improve the classification accuracy.

The promising results encourage us to further combine DSR with the convolutional neural networks (CNN) method which can explore both the spatial and spectral features of HSIs.

References

1. Zhang J, Liu H, Wei Z (2018) Regularized variational dynamic stochastic resonance method for enhancement of dark and low-contrast image. Comput Math Appl 76(4):774–787. https://doi.org/10.1016/j.camwa.2018.05.018

2. Wuehr M, Boerner J, Pradhan C, Decker J, Jahn K, Brandt T, Schniepp R (2018) Stochastic resonance in the human vestibular system-noise-induced facilitation of vestibulospinal reflexes. Brain Stimul 11(2):261–263
3. Benzi R, Sutera A, Vulpiani A (1981) The mechanism of stochastic resonance. J Phys A: Math Gen 14(11):L453
4. Hycza T, Stereńczak K, Bałazy R (2018) Potential use of hyperspectral data to classify forest tree species. N Z J For Sci 48(1):18
5. Shao Z, Yin Z, Song H, Liu W, Li X, Zhu J, Biermann K, Bonilla LL, Grahn HT, Zhang Y (2018) Fast detection of a weak signal by a stochastic resonance induced by a coherence resonance in an excitable GaAs/Al 0.45 Ga 0.55 as superlattice. Phys Rev Lett 121(8):086,806
6. Polak A, Coutts FK, Murray P, Marshall S (2019) Use of hyperspectral imaging for cake moisture and hardness prediction. IET Image Process 13(7):1152–1160
7. Qiao Z, Lei Y, Li N (2019) Applications of stochastic resonance to machinery fault detection: a review and tutorial. Mech Syst Signal Process 122:502–536
8. Qureshi R, Uzair M, Khurshid K, Yan H (2019) Hyperspectral document image processing: applications, challenges and future prospects. Pattern Recognit 90:12–22
9. Xu B, Duan F, Bao R, Li J (2002) Stochastic resonance with tuning system parameters: the application of bistable systems in signal processing. Chaos Solitons Fractals 13(4):633–644
10. Moss F, Ward LM, Sannita WG (2004) Stochastic resonance and sensory information processing: a tutorial and review of application. Clin Neurophysiol 115(2):267–281
11. Benzi R, Parisi G, Sutera A, Vulpiani A (1982) Stochastic resonance in climatic change. Tellus 34(1):10–16
12. Lu H, Li Y, Chen M, Kim H, Serikawa S (2018) Brain intelligence: go beyond artificial intelligence. Mob Netw Appl 23(2):368–375
13. Serikawa S, Lu H (2014) Underwater image dehazing using joint trilateral filter. Comput Electr Eng 40(1):41–50
14. Lu H, Li Y, Uemura T, Kim H, Serikawa S (2018) Low illumination underwater light field images reconstruction using deep convolutional neural networks. Future Gener Comput Syst 82:142–148
15. Lu H, Wang D, Li Y, Li J, Li X, Kim H, Serikawa S, Humar I (2019) Conet: a cognitive ocean network. arXiv:1901.06253
16. Lu H, Li Y, Mu S, Wang D, Kim H, Serikawa S (2017) Motor anomaly detection for unmanned aerial vehicles using reinforcement learning. IEEE Internet Things J 5(4):2315–2322

Generative Image Inpainting

Jiajie Xu, Yi Jiang, Junwu Zhu, and Baoqing Yang

Abstract Recently, image is becoming more and more important as a carrier of information, and the demand of image inpainting is increasing. We present an approach for image inpainting in this paper. The completion model contains one generator and double discriminators. The generator is the architecture of AutoEncoders with skip connection and the discriminators are simple convolutional neural networks architecture. Wasserstein GAN loss is used to ensure our model's stable training. We also give the algorithm of training our model in this paper.

Keywords Image inpainting · Wasserstein GAN · Skip connection · Autoencoders · Convolution neural networks.

1 Introduction

Image inpainting [4] aims to fill missing regions of a damaged image with plausibly synthesized contents. This technology can be widely applied in many fields like medical image restoration, ancient books restoration, and PhotoShop processing. The key problem to be solved in image inpainting is how to make the inpainted image looks real and semantic coherent. To solve this problem, total variation (TV) based approaches [1, 17] and PatchMatch (PM) based method [3] get great success in filling small missing regions, deep learning (DL) based methods [6, 11, 14, 15] are widely adopted to the inpainting task with large missing regions and have made great progress. However, the problem of blurring the boundary between the inpainted regions and the original regions still exists. Beyond that, how to ensure the semantic

J. Xu · Y. Jiang (✉) · J. Zhu · B. Yang
College of Information Engineering, Yangzhou University, Jiangsu, China
e-mail: jiangyi@yzu.edu.cn

J. Zhu
Department of Computer Science and Technology, University of Guelph,
Guelph NIG2K8, Canada

© Springer Nature Switzerland AG 2021
H. Lu (ed.), *Artificial Intelligence and Robotics*,
Studies in Computational Intelligence 917,
https://doi.org/10.1007/978-3-030-56178-9_12

correctness of the inpainted regions is also one of the difficulties in the task of image inpainting.

In this paper, we train a model based on convolution neural networks to fill large-scale missing regions. Our model consists of one generator for generating content to fill the missing regions and double discriminators for discriminating if the inpainted image is visually and semantically plausible. The architecture of the generator of our model is similar to autoencoders [10] with an encoder and a decoder. Different from original autoencoders, our model combines with the characteristics of skip connection [9] and autoencoders. We use skip connection in the generator, which can help our model use the underlying network to enhance the prediction ability of the decoding process, and prevent the gradient vanishing caused by the deep neural network. Similar to the architecture proposed by Iizuka et al. [11], we use the architecture of double discriminators: global discriminator and local discriminator. The difference is that we use the Wasserstein GAN [2] loss to train our model, which can ensure our model's stable training. We demonstrate that the model we proposed is capable of generating realistic and semantically coherent images when inpainting images.

The rest of this paper is structured as follows: In Sect. 2, we discuss multiple image inpainting methods. This is followed in Sect. 3 by details about our model and the training Algorithm of our method is proposed in Sect. 4. Conclusions and future works are given in Sect. 5.

2 Related Work

In order to address these problems of image inpainting, existing methods categorized into two types: one category is texture synthesis methods based on the patch [3], the main idea is to find the boundary of the missing regions to fill the missing regions of the image. The other is methods based on the convolution neural networks (CNNs) [13], the main idea is to extract the features of the image through the deep convolution neural network to understand the image, and then to fill the missing region.

A typical patch-based method is proposed by Barnes [3], which searches for matching patches from the rest regions of the image to fill in missing regions. This method can get results with high quality. But it is limited to some non-semantic image inpainting tasks like background inpainting. It does not perform well in the tasks of semantic image inpainting like face completion, landscape image inpainting. Similarly, other patch-based methods [7] and exemplar-based methods [5, 16, 18] are weak in the task of image inpainting of missing the regions with complex structures. The reason is that the texture synthesis method based on the patch is not enough to get the high characteristics of the image.

The rapid development of deep learning provides new ideas for image inpainting. Deep learning based methods exactly fill the defect of the traditional image inpainting methods, which is lack of semantic coherence and difficult to deal with the problem of large areas or complex structures missing. Neural networks are more powerful to learn high-level semantic information of images and CNNs are effective tools for image processing [12].

More and more researchers use deep learning for image inpainting tasks. One of the first methods based on neural networks is Context-encoder [15], which is proposed by Pathak et al. After this, many methods [11, 14, 19, 20] appear and achieve great success in the job of image inpainting. These methods can generate realistic contents to fill the missing regions. However, these methods sometimes create blurry textures inconsistent with the old regions of the image and do not perform well in the task of image inpainting with complex structures missing. Although the existing methods can repair the damaged image well and generated plausible results, the problem of blurring still exists and is worth studying.

3 Methodology

3.1 Problem Formulation

Assuming x is a complete image and x_m is a partially missing or occluded picture, the job of image inpainting is to find a function f such that inpainted image $f(x_m)$ is as close as possible to the original complete image x. The formal representation is as follows:

$$f = \arg\min_f \| f(x_m) - x \|_2^2 \tag{1}$$

3.2 Framework Overview

This paragraph details the model framework for image inpainting. Given an image with a missing hole, the goal is to fill a missing region of the image so that the entire image is semantically coherent and true. Figure 1 shows our model that consists of one generator and two discriminators.

3.2.1 Generator

To complete the task of image inpainting, the generator of our model uses the architecture of autoencoders with an encoder and a decoder. In this paper, we input the incomplete image into the encoder to encode it into code. Then input the code into the decoder to decode into the inpainted image. In order to avoid gradient vanishing caused by the deep neural networks, skip connection [9] is added between the encoder and the decoder. Skip connection can make sure that the decoding stage can utilize the output of the low-level coding stage of the corresponding resolution to supplement

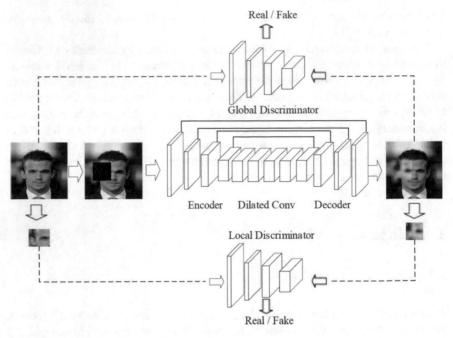

Fig. 1 Network architecture

the decoder with part of the structural feature information lost during the encoder downsampling phase. Also, it can enhance the structure prediction capability of the generator. We use multiple convolution layer architecture in the encoder. Similar to the encoder, the structure of the decoder is symmetric to the encoder. We also use dilated convolution layer instead of fully connected layer between the encoder and the decoder.

3.2.2 Discriminator

The job of the generator is to fill the missing regions of the image. However, one generator is not powerful enough to produce real results. In order to enhance the ability of the generator, we use the discriminator as a binary classifier to discriminate whether the image is from real image dataset or created by the generator. In this paper, we use the architecture of double discriminators with a local discriminator and a global discriminator. Both the local discriminator and the global discriminator are CNNs based architectures. The local discriminator mainly identifies if the generated regions are semantically accurate. The local discriminator can enhance the generating ability of the generator. We use the global discriminator to identify the degree of coherence between the new regions (generated by the generator) and the old regions (original regions). The global discriminator can help the generator produce more semantically consistent results.

3.3 The Joint Loss Function

In this paragraph, we propose the objective function of our model. We first introduce the reconstruction loss L_r. The generator is trained by minimizing L_r. In this paper, we use the L_2 norm loss function instead of the L_1 norm loss function. It is mainly because the L_2 norm penalizes the outliers and is suitable for the task of image inpainting. The reconstruction loss is defined as follows:

$$L_r = \| G(x_m) - x \|_2^2 \tag{2}$$

Due to the use of double discriminators, we apply the Wasserstein GAN loss [2] instead of the original GAN loss [8] in our model. Because the Wasserstein can ensure our model's stable training. The Wasserstein GAN loss is defined as follows:

$$\min_G \max_D V(D, G) = \mathbb{E}_{x \sim p_{\mathbf{data}}(x)}[D(x)] - \mathbb{E}_{z \sim p_z(z)}[D(G(z))] \tag{3}$$

The global discriminate loss and the local discriminate loss are defined as follows:

$$L_{global} = \mathbb{E}_{x_c \sim p_g}[D_g(x_c)] - \mathbb{E}_{x \sim p_{data}}[D_g(x)] \tag{4}$$

$$L_{local} = \mathbb{E}_{m_c \sim p_g}[D_l(m_c)] - \mathbb{E}_{m \sim p_{data}}[D_l(r)] \tag{5}$$

where L_{global} and I_{local} represent the losses of global discriminator and local discriminator respectively. D_g and D_l are the function of global discriminator and local discriminator. x_c is the whole image with generated regions and m_c is regions generated by the generator. x and r are real image and region from real data distribution.

Overall, the total loss function is defined as

$$L = L_r + \lambda_1 L_{local} + \lambda_2 L_{global} \tag{6}$$

where λ_1 and λ_2 are the weights to balance the effects of different losses.

4 Training Algorithm

In this paragraph, we introduce the algorithm of training our model. We use the mini-batch method to mask the images from the dataset in each iteration. Firstly, we sample a mini-batch of images x from training data and mask them with random hole. Then we get a mini-batch of masked images z, real regions before being masked r and masks m. $z = x \odot m$ where $\odot$ represents element-wise multiplication. Then we train the generator s times with L_r loss. After training the generator, we fix the generator and train discriminators t times with L_{global} and L_{local}. Finally, we train the joint model with joint loss L. Input z into the model and output the predicted images c. Combining the masked regions of c with z, we get the final inpainting images $x_i = z + c \odot (1 - m)$.

5   Conclusions and Future Works

In this paper, we propose a novel method base on Generative Adversarial Networks for the task of image inpainting. We use the technology of skip connection in the generator to improve the generating ability of the generator. Double discriminator architecture is used to enhance the prediction ability of our model and make our model create visually and semantically plausibly results. In the future, we will extend our model to deal with the task of complex structures missing. Also, we will optimize our model to complete the task of image denoising and image super-resolution.

Acknowledgements This work was supported in part by the National Natural Science Foundation of China under grant nos. 61872313; the Key Research Projects in Education Informatization in Jiangsu Province under grant 20180012; by the Postgraduate Research and Practice Innovation Program of Jiangsu Province under grant *KYCX* 18_2366; and by Yangzhou Science and Technology under grant YZ201728,YZ2018209; and by Yangzhou University Jiangdu High-end Equipment Engineering Technology Research Institute Open Project under grant YDJD201707.

References

1. Afonso MV, Bioucas-Dias JM, Figueiredo MA (2010) An augmented lagrangian approach to the constrained optimization formulation of imaging inverse problems. IEEE Trans Image Process 20(3):681–695
2. Arjovsky M, Chintala S, Bottou L (2017) Wasserstein gan
3. Barnes C, Shechtman E, Finkelstein A, Dan BG (2009) Patchmatch: a randomized correspondence algorithm for structural image editing. Acm Trans Graph 28(3):1–11
4. Bertalmio M, Sapiro G, Caselles V, Ballester C (2005) Image inpainting. Siggraph 4(9):417–424
5. Criminisi A, Prez P, Toyama K (2003) Object removal by exemplar-based inpainting. In: IEEE Conference on computer cision and pattern recognition
6. Demir U, Unal G (2018) Patch-based image inpainting with generative adversarial networks. arXiv preprint. arXiv:1803.07422
7. Farahnakian F, Liljeberg P, Plosila J (2013) Lircup: linear regression based cpu usage prediction algorithm for live migration of virtual machines in data centers. In: 2013 39th Euromicro conference on software engineering and advanced applications, pp 357–364. IEEE
8. Goodfellow IJ, Pouget-Abadie J, Mirza M, Bing X, Bengio Y (2014) Generative adversarial networks. Adv Neural Inf Process Syst 3:2672–2680
9. He K, Zhang X, Ren S, Sun J (2016) Deep residual learning for image recognition. In: Proceedings of the IEEE conference on computer vision and pattern recognition, pp 770–778
10. Hinton GE, Salakhutdinov RR (2006) Reducing the dimensionality of data with neural networks. Science 313(5786):504–507

11. Iizuka S, Simo-Serra E, Ishikawa H, Iizuka S, Simo-Serra E, Ishikawa H, Iizuka S, Simo-Serra E, Ishikawa H (2017) Globally and locally consistent image completion. Acm Trans Graph 36(4):1–14
12. Krizhevsky A, Sutskever I, Hinton GE (2012) Imagenet classification with deep convolutional neural networks. In: International conference on neural information processing systems
13. LeCun Y, Bottou L, Bengio Y, Haffner P et al (1998) Gradient-based learning applied to document recognition. Proc IEEE 86(11):2278–2324
14. Li Y, Liu S, Yang J, Yang MH (2017) Generative face completion
15. Pathak D, Krahenbuhl P, Donahue J, Darrell T, Efros AA (2016) Context encoders: feature learning by inpainting. In: IEEE conference on computer vision and pattern recognition
16. Qiang Z, He L, Dan X (2017) Exemplar-based pixel by pixel inpainting based on patch shift. In: Ccf Chinese conference on computer vision
17. Shen J, Chan TF (2002) Mathematical models for local nontexture inpaintings. SIAM J Appl Math 62(3):1019–1043
18. Tang XN, Chen JG, Shen CM, Zhang GX (2009) Modified exemplar-based image inpainting algorithm. J. East China Normal Univ
19. Yang C, Lu X, Lin Z, Shechtman E, Wang O, Li H (2017) High-resolution image inpainting using multi-scale neural patch synthesis. In: Proceedings of the IEEE conference on computer vision and pattern recognition, pp 6721–6729
20. Yeh RA, Chen C, Lim TY, Schwing AG, Hasegawajohnson M, Do MN (2016) Semantic image inpainting with deep generative models

Rice Growth Prediction Based on Periodic Growth

Yongzhong Cao, He Zhou, and Bin Li

Abstract On the basis of studying the growth and development characteristics of rice and the historical data of growth in recent years, this paper gives the definition of rice growth quantity, which takes into account the periodic growth and the key growth indexes of each stage, so as to characterize the growth of rice in each growing period. Elman neural network was used to determine the relationship between environmental factors and growth in each growth period. At the same time, in order to improve the convergence speed and accuracy of the algorithm, we propose an adaptive improved genetic algorithm to optimize the forward feedback data. The training samples of the paper are composed of various environmental parameters and physiological indexes of rice at each stage. In the experiment, the network training was carried out with the historical samples of several years to obtain the weights of each layer of the model, and the precision of the improved model was improved.

Keywords Elman neural network · Rice · Growth · Genetic algorithm

1 Introduction

With the rise of agricultural Internet of Things [1], big data [2] and artificial intelligence technology [3], the intelligent management of agricultural production has attracted more and more attention of experts in this field. Among them, rice growth prediction is the key link of agricultural precision management [4, 5]. If a prediction model can be established, the corresponding rice growth trend can be predicted according to the input environmental parameters before actual production, and then the final yield can be estimated, which will have a positive significance in enhancing the potential of rice field production and guiding farming.

Rice growth and development is a complex process of interaction between varieties and environmental factors, so the establishment of its prediction model is also a non-linear and complex problem. At present, there are two different ideas in the

---

Y. Cao (✉) · H. Zhou · B. Li

College of Information Engineering, Yangzhou University, Jiangsu 225000, China

e-mail: caoyz@yzu.edu.cn

© Springer Nature Switzerland AG 2021

H. Lu (ed.), *Artificial Intelligence and Robotics*,

Studies in Computational Intelligence 917,

https://doi.org/10.1007/978-3-030-56178-9_13

field of rice growth prediction: the crop growth model initiated by DeWit [6] in the Netherlands has its own system. They subdivide the crop production system into four different stages, and point out that the laws affecting the growth of each stage are different. They build models around each stage, including HLCROS, BACROS, SUCROS and WODOST, which are based on statistical theory. The law of crop growth does express the general law of crop growth, but the comprehensive model [7, 8] is complex and difficult which is grasped by ordinary people. The other is the model of rice growth prediction based on data analysis [9–11], which excavates the hidden relationship between rice yield and environmental factors such as temperature, light and water. However, it often ignores the different characteristics of influencing factors and indicators of crop growth in each growth period, and its accuracy is limited.

On the basis of previous studies on the influence of environmental factors on rice growth, this paper emphatically analyzed the growth characteristics of rice at different growth stages, and gave the definition of rice growth considering both growth cycle and key growth indicators. Rice growth is a comprehensive index, which is a numerical representation of the growth results in a certain time interval. In this paper, the relationship between environmental factors and rice growth is obtained by using improved Elman neural network, and the growth trend is expressed by the growth value, which provides a new idea for rice growth prediction.

The organization of the paper is as follows:

In Sect. 2, we introduced the overall framework of the rice growth prediction model. In Sect. 3, we describe the definition and mathematical description of rice growth. The fourth section describes the structure of the sample set in the model, and the fifth section proposes improved Elman neural network algorithm. In Sect. 6, the model has been trained, the seventh subsection shows the results of the model training, and in Sect. 8, we summarize the conclusions of the rice prediction model.

2   System Framework of Growth Prediction Model

This prediction system is based on the Agricultural Internet of Things [12, 13]. There are two kinds of data collected through sensor networks and artificial observation. One is the environmental data based on time axis, such as light, water level, temperature and humidity, and the other is the growth index data of rice in each growth cycle. The neural network located in the central server will use this as training samples to get the weight of the prediction network at all levels, and can train the network in real time according to the continuous accumulation of follow-up samples to obtain more accurate prediction model. When it is necessary to predict the growth trend, the system can combine the data of the previous stages of perception of paddy field in this year to predict the growth trend of paddy field layer by layer in the later stage, so as to predict the growth value of rice, and even manually adjust the environmental quantity in the later stage to analyze the growth trend change (Fig. 1).

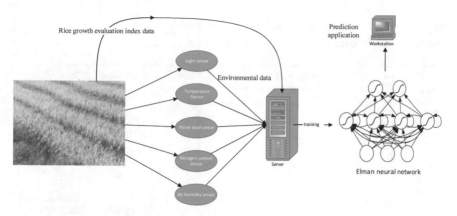

Fig. 1 Rice growth prediction based on Elman neural network

3 Definition of Periodic Growth

In order to evaluate the growth of rice in different growth stages conveniently, the paper defines the descriptive quantity of rice growth as growth amount R. Studies by Zhang et al. [14] show that suitable leaf area index (LAI) is the basis of high yield of rice. The change of leaf area affects and restricts the change of tiller number. The accumulation of dry matter is closely related to these two factors, which together affect the growth of rice.

Generally speaking, the growth of rice can be divided into six periods: Returning green period, Tillering period, Jointing and booting period, Heading and flowering period, Filling period and Maturing period [15–17]. The physiological characteristics of rice plants at different growth stages are different, and each cycle has its own unique physiological characteristics. Therefore, it is necessary to analyze the characteristics of rice at different growth stages in detail. For example, people are mainly concerned about its survival rate in the period of rejuvenation. The number of effective tillers was concerned at Tillering period; The seed setting rate after flowering was mainly considered at Heading and flowering period; The accumulation of dry matter could be examined at Filling period.

Based on this analysis, we divided the different indicators affecting the growth of each period into two categories:

Inheritance indicator is a cumulative measure that acts on all growth cycles of rice. For example, dry matter accumulation is an inheritance index, which is expressed by $ihd_{i,j}$ as the first inheritance index.

Independence Indicator is valid only in a certain growth cycle. For example, the rate of ear formation is an independent indicator of Jointing and booting period. Therefore, it is expressed by $idd_{i,j}$, where i is the first growth cycle and j is the jth observation index of i growth cycle.

In reference to agriculture, ARIMA model [18] and ORYZA2000 model [19] were used to simulate the effects of different physiological characteristics of plant

Table 1 Index quantity of physiological evaluation of growth

Fertility cycle	Reproductive stage	$a_{i,1}$	$a_{i,2}$	$a_{i,3}$	$a_{i,4}$	$a_{i,5}$
Returning green period	1	Dry matter accumulation	Stem tiller number	Leaf area index	Survival rate	Plant height
Tillering period	2				Effective tiller number	Leaf age
Jointing and booting period	3				Earing rate	Clump height
Heading and Flowering Period	4				Setting rate	Ear length
Filling period	5				Grain plumpness	Photosynthetic potential
Maturing period	6				1000-grain weight	Fullness

organs on the final yield potential [20]. According to the characteristics of rice growth stages, two reference quantities were selected as the most important and unique index quantities for each growth stage. As shown in Table 1, the selected births for each growth cycle were selected in the table. Rational character index.

The growth situation of rice depends to some extent on the upper limit of the index of yield factors. The growth trend of each growth period is interlinked in the process of rice growth. The growth trend of any period can not be ignored. It directly or indirectly determines the final yield of rice.

Growth is a comprehensive index. According to the division of growth cycle, a numerical value is used to characterize the growth trend in a certain time interval. The higher the value is in the time interval, the better the growth trend will be in this stage.

The paper divides the evaluation layer according to the agronomic growth period. One growth period is an evaluation layer, and there are six evaluation layers. The evaluation elements of different evaluation layers are not identical. In this paper, each evaluation element matrix is regarded as the evaluation element matrix of each period. The formula is as follows:

$$G = \begin{pmatrix} a_{1,1} & \cdots & a_{1,j} \\ \vdots & \ddots & \vdots \\ a_{i,1} & \cdots & a_{i,j} \end{pmatrix} i \in [1,6], j \in [1,n] a_{i,j} = \begin{cases} idd_{i,j} \, Independece\,indicator \\ ihd_{i,j} \, Inheritance\,indicator \end{cases}$$

(1)

In the formula, i and j are positive integers, i is the stage of rice current growth period, j is the serial number of rice growth evaluation index parameters, and j is the value of the jth rice evaluation index in the stage of rice growth.

If the weight value of the jth rice evaluation index parameter in the first growth stage is set as the weight value, then the weight matrix corresponding to the evaluation index in each cycle is set as.

$$W = \begin{bmatrix} w_{1,1} & \cdots & w_{i,1} \\ \vdots & \ddots & \vdots \\ w_{j,1} & \cdots & w_{j,i} \end{bmatrix} \quad i \in [1, 6], j \in [1, n] \tag{2}$$

Then the principal diagonal matrix of the product of the factor matrix G and the weight matrix W corresponding to rice growth evaluation is.

$$R' = GW = \begin{bmatrix} a_{1,1} & \cdots & a_{1,j} \\ \vdots & \ddots & \vdots \\ a_{i,1} & \cdots & a_{i,j} \end{bmatrix} \begin{bmatrix} w_{1,1} & \cdots & w_{i,1} \\ \vdots & \ddots & \vdots \\ w_{j,1} & \cdots & w_{j,i} \end{bmatrix} \tag{3}$$

For formal neatness, we multiply a unit matrix here, take the principal diagonal line and multiply it by a full 1 matrix E of row i to get the growth matrix R.

$$R = R'E = \begin{bmatrix} R'_1 & 0 & 0 \\ 0 & \ddots & 0 \\ 0 & 0 & R'_i \end{bmatrix} \begin{bmatrix} 1 \\ \vdots \\ 1 \end{bmatrix} = \begin{bmatrix} R'_1 \\ \vdots \\ R'_i \end{bmatrix} \tag{4}$$

In order to facilitate comparison and normalization into an indicator system, the nearer to one, the better the overall growth of the plant. On the contrary, the nearer to zero, the worse the growth.

4   Two Data Sets of the Model

In the prediction model, we define two data sets. The first data set is the environmental data of each growth period, which comes from the real-time collection of sensor nodes distributed in paddy fields. The other data set is the collection of growth data indicators corresponding to each growth period. Some sources and the field are specially equipped with sensors, most of which come from the actual manual statistics and measurement in the field. We store it in standard XML form.

Definition 1 Sample Sequence Number Set ($ID_{envSample}$), Unique identity, $ID_{envSample} = \{$Year, PlantingSeason, BlockNumber, GrowthStage$\}$.

$Year \in [2008, 2019]$, $PlantingSeason \in [01, 03]$, $BlockNumber \in [0001, 9999]$, $GrowthStage \in [1, 6]$.

Definition 2 Environmental data sets are collections of collected data, E = {DayVar, TemperatureVar, lightHourVar, WindSpeedVar, WaterHeightVar, NitrogenVar}. TemperatureVar represents average temperature, lightHourVar Represents the duration of illumination, WaterHeightVar represents the water level of the ground to the surface, NitrogenVar represents the nitrogen content in the soil.

The format of the environmental factor data set is as follows:

<RECORD>

    <id> $ID_{envSample}$ </id>

    <Period> PeriodVar</Period>

    <GrowingDays>DayVar</ GrowingDays >

    <AvgTemperature>TempVar</AvgTemperature>

    <DaylightHour>LightHourVar</DaylightHour>

    <AvgWindSpeed>WindSpeedVar</AvgWindSpeed>

    <WaterHeight> WaterHeightVar</WaterHeight>

    <NitrogenContent>NitrogenVar</NitrogenContent>

</RECORD>

At the end of each stage, the environmental indicators correspond to a growth evaluation index in the respective cycles, so the number of the evaluation indicator data set is one-to-one correspondence with the number in the environmental data set above $ID_{evaluation} = ID_{envSample}$.

Definition 3 The corresponding growth evaluation indicators in the six growth cycles of rice are not identical, so it is necessary to distinguish different growth stages, as shown in Table 1 $Q = \{ihd_{i,1}, ihd_{i,2}, ihd_{i,3}, idd_{i,1}, idd_{i,2}\}$.

Take the format of the evaluation indicator data set of the greening period as an example, as follows:

<RECORD>

    <id> $ID_{evaluation}$ </id>

    <period> PeriodVar </period>

    <Drymatteraccumulation>DryMatterAccVar</Drymatteraccumulation>

    <Tillersofnumber>TillersofNumVar</Tillersofnumber>

    <LAI>LAIVar</LAI>

    <Survivalrate>SurvivalrateVar</Survivalrate>

    <Plantheight>PlantHvar</Plantheight>

</RECORD>

In this evaluation indicator data set, there are two types of indicators, three of which are inherited indicators, as follows:

$$\{DryMatter\,AccVar,\,Tillersof\,Num,\,LAIVar\} \in ihd_{i,j}.$$

The remaining two indicators are the independence indicators for each stage, as follows:

$$\{Survivalrate,\,Plant\,HVar\} \in idd_{i,j}.$$

5 Improved Circulating Neural Network Based on Genetic Algorithms

The difficulty of establishing rice growth prediction model lies in how to select suitable training methods to train the samples. There are many methods to predict the growth trend by using neural network [21, 22], such as BP [23], RBF [24], GRNN [25], etc. The BP neural network is considered first. It is a very classical pre-feedback neural network with simple structure, easy to understand algorithm and good versatility. Through information forward propagation and error back feedback, the final result is approached, but it is easy to fall into local extreme points in the training process, and when the network structure is too large, it is easy to appear "over-fitting" phenomenon. Rice growth is a dynamic process with time, so choosing a dynamic neural network will be more in line with the needs of this study. Elman cyclic neural network [26–28] is selected here. By adding an internal feedback, the network has the ability to adapt to time-varying characteristics. The forward feedback data are optimized and screened by genetic algorithm, so as to accelerate the convergence speed of the network.

The evaluation methods of growth period are different and the growth trend of the previous cycle directly affects the growth trend of the next cycle. It is reflected in the model that six growth period models need to be trained successively, and the growth prediction models of Returning green period stage, Tillering period, Jointing and booting period, Heading and flowering period, Filling period and Maturing period are generated respectively. Maturity growth prediction model, record $M_{ReturningGreen}$, $M_{tillering}$, $M_{JointingBooting}$, $M_{HeadingFlowering}$, $M_{Filling}$, $M_{Maturing}$.

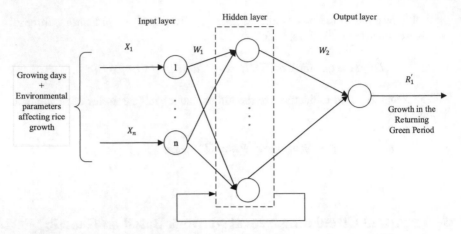

Fig. 2 The network structure of the green-back growth model

In the first stage, there is no previous cycle, and all of them are not affected by the growth trend of the previous cycle. Therefore, the six staged models can be divided into two categories according to the input–output structure. That is to say, $M_{ReturningGreen}$ is for one category, the rest $M_{\text{tillering}}$, $M_{\text{JointingBooting}}$, $M_{\text{HeadingFlowering}}$, M_{Filling}, M_{Maturing} are for another category. Firstly, as shown in Fig. 2, a growth model is established and trained, and the weight files of several environmental factors on the evaluation index of the growth in the green-back period were obtained. In Fig. 2, considering the correlation between inheritance index and input layer, internal feedback of inheritance index screened by genetic algorithm is added.

As shown in Fig. 3, $n = 6$, $i \in [2, 6]$, the weight files of multiple environmental factors for other multiple growth stages were obtained.

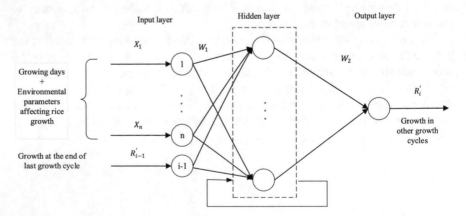

Fig. 3 Schematic diagram of other periodic network structures

The weight file above can not directly get the value of growth, so we need to know the relationship between the value of growth and its evaluation index. The weight of each index on growth is given dynamically by agricultural experts. For example, in heading and flowering period, through consulting relevant experts, the proportion of knowing seed setting rate is generally about 0.27, and the leaf area is in the current period. Not very sensitive, accounting for about 0.15. Through this empirical average, the corresponding growth R_i' can be obtained by formula (5).

$$R_i' = \sum_{j=1}^{n} \lambda_{ij} A_i \ i \in [1, 6], \ j \in [1, 5] \tag{5}$$

In this formula, A_i is the data of physiological traits in different growth stages of rice and λ_{ij} is the weight ratio of the jth physiological index in the 1st growth stage to the current growth volume were used.

As for the algorithm of genetic algorithm [29], we only give the design of different coding methods [30]. Because the growth indicators are divided into inheritance indicators and independence indicators, the correlation of inheritance indicators should be stronger than that of independence indicators. If they are equated, the growth characteristics of rice can not be taken into account, so the proportion of inheritance indicators in the gene is increased. In addition, the network weight of rice growth prediction model is mostly decimal. If the length of individual coding is long and the space is large, the calculation of fitness and the operation of individual crossover and mutation increase the complexity. Considering the mapping error of continuous function discretization in binary coding and the disadvantage of expressing rice growth, in order to facilitate the search of genetic algorithm in complex space. With the construction, real number coding is used here, each individual is a real number string. The specific algorithm is described as follows:

Algorithm 1   Improved genetic algorithm optimization based on Elman neural network for rice growth prediction model

Input: Training forecast data $Data$; Population size $Size$; Genetic algebra $Maxgen$;   Minimum target error $MinError$; The number of input layer, hidden layer, acceptance layer, and output layer nodes are respectively In、 H_1、 H_2、 O_m;

Output: Initial Values of Weights and Thresholds of Elman Neural Network Algorithms in Rice Growth Prediction Models W_1、 W_2、 W_3、 θ_1、 θ_2;

Encoding a chromosome with a real number encoding;

The first generation ($Gen=1$) chromosomes were randomly generated by controlling the proportion of inheritance indices, in which the population size was $Size$

For i=1: $Size$

Decoding each chromosome in the population and substituting into the Elman neural network;

$Data$ input Elman neural network;

Calculate the prediction $Error$;

Calculate individual fitness $Fitness(j)$ according to formula $F = \dfrac{1}{\sum_{i=1}^{k}[y_d(k)-y(k)]^T[y_d(k)-y(k)]+\zeta}$;

End (y_d For the actual output value of the network, y is the expected output, avoiding the denominator being 0, ζ which is a smaller constant.)

For $Gen = 1$: $Maxgen$

Selection;

Adaptive crossover;

Adaptive Mutation;

For k=1: Size

Decoding chromosomes in population and substituting them into Elman neural network

Data input Elman neural network

Calculating prediction $Error$

Calculate individual fitness $Fitness(j)$ according to formula $F = \dfrac{1}{\sum_{i=1}^{k}[y_d(k)-y(k)]^T[y_d(k)-y(k)]+\zeta}$;

End

Compare the fitness values and find the individual $Fitness_{max}$ with the highest fitness value and the individual with the least fitness $Fitness_{min}$;

Population update, replacing individual $Fitness_{max}$ with individual $Fitness_{min}$;

Updating Optimal Individual Records;

$Gen = Gen + 1$;

End

Decode the best individual and get the output

---

6 Model Training

There may be disturbance anomalies in the environmental data collected by various sensors in the experimental field, and the scalar units of environmental factors indicators collected lead to large differences among the data, which is not in an order of magnitude. It is difficult to obtain ideal prediction results if the network training prediction is conducted directly as input. In practical applications, the integrity and validity of data are extremely important, so it is generally necessary to normalize the data [31].

(1) Normalization. The units of data collected from paddy fields are not uniform, such as light and temperature units. The values are too different to be used directly, and the data fluctuate too much. In order to improve the generalization ability of the neural network, the input is normalized to the [0, 1] interval, but the log-Sigmiod function changes very slowly in the [0, 0.1] and [0.9, 1] intervals, which is not conducive to feature extraction. In order to improve the efficiency, the input is normalized to the [0.1, 0.9].

The treatment formula is.

$$X_{di} = \frac{X_i - X_{min}}{X_{max} - X_{min}} + 0.1 \tag{6}$$

Formula: X_{di} is the processed data, the maximum value is X_{max}, the minimum value is X_{min}.

(2) Setting of input and output parameters

The input and output of each model are shown in Table 2 as follows.

The relationship between R_i' and M_i is one-to-one. i is the number of growth stages, $i \in [1, 6]$. Taking the grain filling stage as an example, the growth of rice in the fifth growth cycle of rice is predicted by the rice growth prediction model ($M_{Filling}$).

(3) Number of Hidden Layers and Nodes

Table 2 Input and output parameters of each model

Rice growth model	Input parameters	Output parameters
Growth prediction model in Returning Green Period $M_{ReturningGreen}$	Growing days	Rice growth in returning green period R_i'
	Environmental factor values collected	
Growth prediction model in other growth Period ($M_{tillering}$, $M_{JointingBooting}$, $M_{HeadingFlowering}$, $M_{Filling}$, $M_{Maturing}$)	Growing days	Rice growth in different growth cycles R_{i-1}'
	Environmental factor values collected	
	Growth at the end of last growth period R_{i-1}'	

The speed and accuracy of learning will be affected by the number of hidden layers and nodes. If there are too many points, the training time will be greatly increased, and there may be problems of fitting. On the contrary, there will be relatively little information and the training effect is not ideal. The basic principle of determining the hidden layer is to use the network structure as compact as possible to meet the accuracy requirements [32]. Therefore, the number of hidden layers and nodes must be within a reasonable range, so that the training accuracy is high and the number of training times is reduced as much as possible. The number of nodes is related to the number of input and output nodes as well as the complexity of the problem. Here, the following methods are used: first, the empirical formula is roughly ranged, and then the best value is obtained through experiments.

$$P = L + \sqrt{M + N} \tag{7}$$

In the formula, P, M and N are the number of nodes in the hidden layer, input layer and output layer respectively, and L is a positive integer and $L \in [1, 10]$. When calculating the number of hidden layer nodes of $M_{ReturningGreen}$, we try to set the number of hidden layer nodes to be 5, 6, ... 12, 13 enter the simulation experiment, observe the effect of the experiment, and finally determine the appropriate number of hidden layer nodes.

Taking the samples in the rejuvenation period as an example, the experiment was carried out on MATLAB, in which the training steps were set to 4000 steps and the error value was set to 0.00001.

Table 3 shows that the convergence accuracy of the network is the best when the number of hidden layer nodes is 8. So the number of nodes selected is 8. Similarly, five growth stages models, such as tillering stage, are also calculated by this method. The results show that 9 is the best time, so the number of hidden layer nodes of five growth stages models, such as tillering stage, is 9, and the number of hidden layer nodes of the whole model is 8.

(4) Number of nodes in the undertaking layer

The number of nodes in the receiving layer should be the same as that in the hidden layer depending on the characteristics of the network. The number of nodes in the receiving layer corresponding to each model is shown in Table 4.

Table 3 Number of hidden layer nodes and total errors in neural network simulation

Number of nodes in hidden layer	Total error	Number of nodes in hidden layer	Total error
4	9.48e−04	9	4.37e−04
5	8.64e−04	10	9.82e−05
6	7.23e−04	11	2.27e−04
7	2.56e−04	12	1.37e−04
8	7.73e−05	13	2.42e−04

Table 4 Number of nodes in succession layer of each model

	$M_{ReturningGreen}$	$M_{\text{tillering}}$, $M_{\text{JointingBooting}}$, $M_{\text{HeadingFlowering}}$, M_{Filling}, M_{Maturing}
Number of nodes in the acceptance layer	8	9

Neural network is a data-driven solution. The reliability of the samples is very important. All the samples in this paper come from the measured data of the rice irrigation experimental field in Agricultural College. The collected sample data are randomly divided into three parts: training sample, test sample and test sample. 80% of the data were used as training data and 20% as validation data.

The experimental algorithm is based on Python 3. Some key codes are attached.

```
dataset = pd.read_csv('./evn_input/env1.csv', index_col=None)

env1_outputdata = dataset['Ingrowth']

output_env1 = env1_outputdata.values

env1_inputdata = dataset[['AvgTemperature', 'DaylightHour', 'AvgWindSpeed','WaterHeight','NitrogenContent']]

input_env1 = env1_inputdata.values

X_train, X_test, Y_train, Y_test = train_test_split(input_env1, output_env1, test_size=0.4, random_state=20)

clf = MLPRegressor(solver='sgd',momentum=0.8, alpha=1e-4,

                   hidden_layer_sizes=(12,),

                   learning_rate_init=1e-5, max_iter=10000,

                   random_state=32).fit(X_train, Y_train)

sc = clf.score(X_test, Y_test)

print(sc)
```

7 Test Result

In order to illustrate the validity of the improved Elman neural network in rice growth prediction model, the following three models were used to predict rice growth.

(1)   Traditional BP Neural Network
(2)   Standard Elman Neural Network
(3)   Improved Elman Neural Network.

The prediction results of the three kinds of neural networks are shown in Fig. 5. The predicted values of the traditional BP neural network and the actual data have low fitting degree, some inaccurate points exist, and the predicted values of the standard Elman neural network have improved in the accuracy of the algorithm, but there is

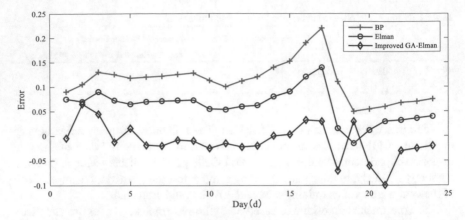

Fig. 5 Prediction error diagrams of three kinds of neural networks

still a reverse trend of individual error points, which indicates that the Elman neural network will fall into the local extreme point and pass through the genetic algorithm. The fitting degree of Elman improved by the algorithm is very high, and the error between the predicted value and the real value is smaller. It can be seen that the improved Elman neural network has better predictive value in the prediction model of rice growth.

In order to verify the performance of the improved Elman neural network, the Elman neural network and the improved Elman neural network are trained respectively, and their curves are compared, as shown in Fig. 6.

The precision of traditional BP neural network, standard Elman neural network and improved Elman neural network in rice growth prediction was compared and analyzed by means of mean square deviation (MSE) and mean absolute error (MAE), as shown in Table 5.

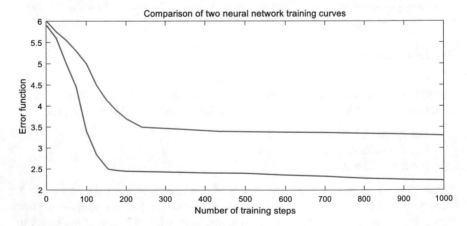

Fig. 6 Performance curve of model training process

Table 5 Comparisons of algorithmic errors

Neural network type	Average MSE	Average MAE
BP neural network	0.01710	0.15423
Elman neural network	0.00682	0.08440
Improved Elman neural network	0.00212	0.02603

From the point of error analysis, as shown in Table 5, compared with the traditional BP and Elman neural networks, the improved algorithm has smaller error range, smaller points with larger prediction error and smaller fluctuation, which greatly improves the prediction accuracy. From the point of convergence speed, as shown in Fig. 6, the improved algorithm gradually stabilizes after 150 steps, while Elman neural network takes about 250 steps to complete the convergence, which shows that the convergence speed of the improved Elman neural network is faster than Elman. Therefore, the prediction effect of improved Elman neural network in rice growth prediction model is better than that of standard Elman neural network and traditional BP neural network.

8 Conclusion

Rice growth is a relatively complex dynamic process, which is affected by many environmental factors such as light, temperature, water and so on. In order to accurately predict the growth of rice in different growth stages, on the basis of previous studies, this paper deeply studies the quantitative evaluation index of each growth stage, creatively puts forward the concept of rice growth amount to represent the growth of different growth stages, and models different growth stages separately. A rice growth model based on improved Elman neural network is proposed, which realizes the Non-Proliferation of rice. Prediction function of rice growth at the same growth stage.

References

1. Zhao Y, Cao N (2017) Research on traceability of agricultural products based on internet of things. In: 2017 IEEE international conference on computational science and engineering (CSE) and IEEE international conference on embedded and ubiquitous computing (EUC). IEEE
2. Wolfert S, Ge L, Verdouw C et al (2017) Big data in smart farming—a review. Agric Syst 153:69–80
3. Lu H, Li Y, Chen M et al (2017) Brain intelligence: go beyond artificial intelligence. Mobile Netw Appl
4. Huang Y, Hoffmann WC, Lan Y et al (2009) Development of a spray system for an unmanned aerial vehicle platform. Appl Eng Agric 25(6):803–809

5. Mekala MS, Viswanathan P (2017) A novel technology for smart agriculture based on IoT with cloud computing. In: 2017 IEEE international conference on I-SMAC (IoT in social, mobile, analytics and cloud) (I-SMAC). IEEE

6. de Wit CT (1965) Photosynthesis of leaf canopies. Agric Res Rep. Centre for Agricultural Publication and Documentation (PUDOC), Wageningen

7. Lin Z, Mo X, Xiang Y (2003) Review of crop growth models. J Crops 29(5):750–758

8. Lecerf R, Ceglar A, López-Lozano R et al (2018) Assessing the information in crop model and meteorological indicators to forecast crop yield over Europe. Agri Syst S0308521X17310223

9. Gandhi N, Petkar O, Armstrong LJ (2016) Rice crop yield prediction using artificial neural networks. In: Technological innovations in Ict for agriculture & rural development. IEEE

10. Kulkarni S, Mandal SN, Sharma GS et al (2018) Predictive analysis to improve crop yield using a neural network model. In: 2018 international conference on advances in computing, communications and informatics (ICACCI). IEEE, pp 74–79

11. Cao Y, Zhu J, Guo Y et al (2018) Process mining-based medical program evolution. Comput Electr Eng 68:204–214

12. Lu H, Wang D, Li Y et al (2019) CONet: a cognitive ocean network

13. Jiang R, Zhang Y (2013) Research of agricultural information service platform based on internet of things. In: 2013 12th international symposium on distributed computing and applications to business, engineering & science. IEEE, pp 176–180

14. Zhang L, Su Z, Zhang Y (2004) Study on the relationship between stem structure and leaf area index and yield of rice at jointing stage. J Yangzhou Univ (Agri Life Sci) 25(01):55–58

15. Peng S, Huang J, Sheehy JE et al (2004) Rice yields decline with higher night temperature from global warming. Proc Natl Acad Sci U S A 101(27):9971–9975

16. Yu HY, Wang X, Li F et al (2017) Arsenic mobility and bioavailability in paddy soil under iron compound amendments at different growth stages of rice. Environ Pollut 224:136–147

17. Wu C, Lu L, Yang X et al (2010) Uptake, translocation, and remobilization of zinc absorbed at different growth stages by rice genotypes of different Zn densities. J Agri Food Chem 58(11):6767–6773

18. Cai C, Yang C, Mo H et al (2018) Prediction and analysis of rice yield in China based on ARIMA model. Hybrid Rice 33(2):62–66

19. Li T, Angeles O, Marcaida M et al (2017) From ORYZA2000 to ORYZA (v3): an improved simulation model for rice in drought and nitrogen-deficient environments. Agric For Meteorol 237–238:246–256

20. Nisar S, Arora VK (2018) Analysing dry-seeded rice responses to planting time and irrigation regimes in a subtropical environment using ORYZA2000 model. Agri Res

21. Serikawa S, Lu H (2014) Underwater image dehazing using joint trilateral filter. Comput Electr Eng 40(1):41–50

22. Lu H, Li Y, Uemura T, Kim H, Serikawa S (2018) Low illumination underwater light field images reconstruction using deep convolutional neural networks. Future Gener Comput Syst 82:142–148

23. Liu M, Liu X, Li M et al (2010) Neural-network model for estimating leaf chlorophyll concentration in rice under stress from heavy metals using four spectral indices. Biosys Eng 106(3):223–233

24. Rahimzadeh H, Sadeghi M, Ghasemi-Varnamkhasti M et al (2019) On the feasibility of metal oxide gas sensor based electronic nose software modification to characterize rice ageing during storage. J Food Eng 245:1–10

25. Dong J, Dai W, Xu J et al (2016) Spectral estimation model construction of heavy metals in mining reclamation areas. Int J Environ Res Public Health 13(7):640

26. Wang J, Zhang W, Li Y et al (2014) Forecasting wind speed using empirical mode decomposition and Elman neural network. Appl Soft Comput 23(Complete):452–459

27. Yang J, Honavar V (2002) Feature subset selection using a genetic algorithm. IEEE Intell Syst Appl 13(2):44–49

28. Lu H, Li Y, Mu S, Wang D, Kim H, Serikawa S (2018) Motor anomaly detection for unmanned aerial vehicles using reinforcement learning. IEEE Internet of Things J 5(4):2315–2322

29. Canca D, Barrena E (2018) The integrated rolling stock circulation and depot location problem in railway rapid transit systems. Transport Res Part E Logist Transport Rev 109:115–138
30. Nemati M, Braun et al. Optimization of unit commitment and economic dispatch in microgrids based on genetic algorithm and mixed integer linear programming. Appl Energy 210
31. Huang W, Oh SK, Pedrycz W (2016) Polynomial neural network classifiers based on data preprocessing and space search optimization. In: Joint international conference on soft computing & intelligent systems. IEEE
32. Chiliang Z, Tao H, Yingda G et al (2019) Accelerating convolutional neural networks with dynamic channel pruning. In: 2019 data compression conference (DCC). IEEE, p 563

Yongzhong Cao is currently working as a professor of the College of Information Engineering at Yangzhou University. His main research areas are workflow technology and service compute.

He Zhou is a postgraduate student at Yangzhou University. His main research interests are Artificial Intelligence and Internet of things.

Bin Li is currently working as a professer of the College of Information Engineering at Yangzhou University. His main research areas are Artificial Intelligence, software engineering.

A Multi-scale Progressive Method of Image Super-Resolution

Surong Ying, Shixi Fan, and Hongpeng Wang

Abstract In recent year, researchers have gradually focused on single image super-resolution for large scale factors. Single image contains scarce high-frequency details, which is insufficient to reconstruct high-resolution image. To address this problem, we propose a multi-scale progressive image super-resolution reconstruction network (MSPN) based on the asymmetric Laplacian pyramid structure. Our proposed network allows us to separate the difficult problem into several subproblems for better performance. Specially, we propose an improved multi-scale feature extraction block (MSFB) to widen our proposed network and achieve deeper and more effective feature information exploitation. Moreover, weight normalization is applied into MSFB to tackle the gradient vanishing and gradient exploding problem, and to accelerate the convergence speed of training. In addition, we introduce pyramid pooling layer into the upsampling module to further enhance the image reconstruction performance by aggregating local and global context information. Extensive evaluations on benchmark datasets show that our proposed algorithm gains great performance against the state-of-the-art methods in terms of accuracy and visual effect.

Keywords Singe image super-resolution · Multi-scale progressive network · Asymmetric laplacian pyramid · Pyramid pooling layer · Weight normalization

S. Ying (✉) · S. Fan · H. Wang
Harbin Institute of Technology (Shenzhen), Shenzhen, Guangdong, China
e-mail: yingsurong@stu.hit.edu.cn

S. Fan
e-mail: fanshixi@hit.edu.cn

H. Wang
e-mail: wanghp@hit.edu.cn

© Springer Nature Switzerland AG 2021
H. Lu (ed.), *Artificial Intelligence and Robotics*,
Studies in Computational Intelligence 917,
https://doi.org/10.1007/978-3-030-56178-9_14

1   Introduction

Super-resolution (SR) image reconstruction technology generally refers to a computer vision task that recovers a high-resolution (HR) image with sharp edges, reasonable details and clear contours through single or multiple low-resolution (LR) images. In this paper, we focus on single image super-resolution (SISR), which reconstructs the corresponding high-resolution image from single low-resolution input image. High-resolution images can provide a wealth of detailed information that is indispensable in many applications. At present, SISR has been widely applied in the field of medical images [1], public security [2, 3], remote sensing [4], and high-definition television. SISR has also become the basis of many computer vision applications such as face recognition, image restoration, target recognition and tracking.

With the development of convolutional neural networks (CNN) and deep learning, many related methods have been applied into image super-resolution. Compared with traditional image super-resolution algorithms, deep learning based methods have shown a significant advantage. Among them, Dong et al. [5] firstly proposed a three-layer CNN-based image super-resolution reconstruction algorithm (SRCNN). Consequently, deep learning based methods have attracted extensive attention. Many networks are proposed to acquire higher accuracy by continually increasing the depth of networks, such as VDSR [6], DRCN [7], and DRRN [8]. As the network depth grows, the number of parameters increases. The oversized networks will suffer from the gradient vanishing and gradient exploding problem. To prevent from it, recursive learning, residual learning and skip connection methods are gradually applied into image SR algorithm. Recent studies have shown that many networks such as SRDenseNet [9], EDSR [10] and RDN [11] fully use these structures to address the gradient problem and to obtain better performance. However, the direct image reconstruction structure employed in these methods, which performs the upsampling operation in one-step, greatly increases the learning difficulty of larger scale factors (such as 4 times or 8 times). In addition, different networks are required for different scale factors. In order to address these drawbacks, Wei-Sheng Lai et al. [12, 13] proposed a new image reconstruction structure, namely the deep Laplacian pyramid SR network (LapSRN), which consists of a cascade of convolution neural networks to achieve a progressive reconstruction process of HR image. This structure decomposes a difficult problem into multiple simpler subproblems, reducing the learning difficulty and obtaining better performance. Besides, it occupies less storage resources than the direct reconstruction structure, especially for larger scale factors.

The SR problem usually assumes that LR image is acquired from HR image through the low-pass filtering degradation or the downsampling degradation. Since the irreversible degradation will cause the loss of high-frequency information of HR images, it leads to a phenomenon that multiple HR images may result in an identical LR image. Thus, the image SR reconstruction is actually a typical ill-posed problem. To tackle this problem, many researchers have conducted plenty of effective methods. But these methods have certain limitations:

(1) Most of algorithms employ one-step image reconstruction by pre- or post-unsampling structure. Different scale factors images have to require for different reconstruction networks, while progressive reconstruction structure can produce SR images with different scale factors by single network.

(2) Many deep learning based algorithms use mean square error (MSE) as loss function. However, those methods with MSE are essentially an average result of multiple HR images that recovered from an identical LR image. Therefore, using MSE as loss function is no good for finding the potential multi-modal distribution between LR-HR image pairs. What's worse, it makes the recovered image lose its high-frequency details and results in poor visual effects and too smooth image edges.

(3) Many networks have proposed numerous methods to address the gradient vanishing and gradient exploding problem, including recursive learning, residual learning and skip connection. However, some low-level features will still gradually disappear in the information flows, causing high-level networks impossible receiving abundant features.

(4) Many existing networks constantly increase the depth of networks to improve the reconstruction performance. However, these methods ignore the importance of LR image features, and underutilize feature information extracted by low-level networks.

Compared with the direct reconstruction structure, the progressive reconstruction structure [14] gains better performance, especially for large scale factors. It not only produces visually high-quality images, but also reconstructs the output SR images closer to real images. Motivated by these superiority, we propose a multi-scale progressive image super-resolution reconstruction network (MSPN) to extract abundant local features and to improve low-level representation.

We train our networks on DIV2K [15] dataset and evaluate it against many other methods on a variety of testing datasets, where our proposed network shows improved performance in terms of PSNR and SSIM, especially at larger upsampling scale factors.

In summary, our main contributions of this paper are as follows:

- An improved multi-scale feature extraction block (MSFB) is proposed to enhance image feature characterization. MSFB widens the network by adopting multiple filters with different kernel sizes on the same level.
- To alleviate the training difficulty of network and to address the gradient vanishing and gradient exploding problem, weight normalization is adopted into MSFB.
- We apply pyramid pooling layer into the upsampling module to incorporate the context information of different subregions, and to further improve the capability of obtaining global information. Pyramid pooling layer can fully exploit local and global context information to achieve more competitive image reconstruction performance.

The remainder of this paper is organized as follows. Section 2 introduces the related work of image super-resolution task. Section 3 gives an overview of our proposed network MSPN and describes the details about network architecture. We present the evaluation results and make comparisons with other state-of-the-art algorithms quantitatively and qualitatively in Sect. 4. Section 5 concludes this paper.

2 Related Works

Image SR has been developed for more than one decade, and researchers have proposed numerous innovative methods, including traditional image SR algorithms [16, 27, 29], such as interpolation-based, reconstruction model-based, traditional machine learning based methods, and deep learning based image SR algorithms [5–14, 17–19]. Among them, SRCNN [5] proposed by Dong et al. firstly introduced deep learning into image SR reconstruction, and LapSRN [12, 13] proposed by Lai et al. firstly introduced progressive reconstruction structure into image SR.

2.1 Progressive Image Super-Resolution Reconstruction

Different from direct reconstruction structure that performs upsampling operation in one-step, progressive reconstruction gradually increases the scale factor step by step to produce the output SR image. It effectively reduces the computational complexity and avoids the visible reconstruction artifacts.

The Laplacian super-resolution network (LapSRN) proposed by Lai et al. [12, 13] firstly introduced the Laplacian pyramid structure into image SR reconstruction. Based on the multi-cascaded structure, LapSRN achieves a progressive image SR reconstruction by decomposing a large scale factor into multiple smaller scale factors. It can acquire effective information from intermediate pyramid layers and recover the output SR images with different scale factors through single network.

Adopting the progressive reconstruction structure, Wang et al. [14] proposed a fully progressive method of image SR (proSR). It combines progressive reconstruction ideas with network architecture and network training respectively. The architecture principle of proSR network is similar to LapSRN, except that it improves the original architecture of pyramid network from some aspects. Firstly, it proposes asymmetric pyramid architecture to pay more attention on LR space. Secondly, it simplifies the way of information flows to improve the reconstruction performance and convergence speed of training. Finally, it adopts a curriculum-learning training strategy, where the image SR reconstruction task is designed as an easy-to-hard training process.

2.2 Weight Normalization

In machine learning, the input data usually needs to satisfy an assumption of independent and identical distribution (IID), while this assumption is difficult to be implemented in the deep neural network. Because when it refers to multi-layer neural network, the parameter update of each layer will change the distribution of the input data of its upper layer. And after stacking more layers, the distribution will change sharply, making high-level networks have to constantly readjust itself to the updated low-level data.

To tackle this problem, the normalization method is applied into the deep neural network with various transformation methods. It makes the input data of each layer approximately satisfy with the assumption of IID, and limits the output of each layer within a certain range to prepare for succeeding operations. Common normalization methods include batch normalization (BN) [20], layer normalization (LN), weight normalization (WN) [21], and group normalization.

Many previous works have found that BN is not applicable to train image SR network, because of mini-batch dependency, strong regularization side-effects, and requirement for different formulations in training and testing process. However, as the depth of neural networks increases, lack of normalization operations makes networks difficult to be trained.

Yu et al. [22] proposed that adopting WN into networks achieved higher accuracy than BN or without any normalization during the training or testing process. By normalizing the weight vector, WN speeds up the convergence of training, improves the image reconstruction performance and enables networks to be trained at higher learning rates. So, we adopt weight normalization into our proposed network to improve the stability of the training process.

2.3 Pyramid Pooling

In deep neural networks, the size of the receptive fields represents the amount of contextual information it contains. Moreover, as the depth of networks increases, the size of the receptive fields also increases. However, in image processing, the size of the receptive fields is usually insufficient to receive all the global context information, resulting in poor image quality. In order to address this problem, there exist two common strategies: one is to use the global average pooling method, but this method may cause networks to lose the spatial relationship between context information and to blur the reconstructed image; the other is to smoothly concatenate features from different levels, which can reduce the information loss among different subregions.

Zhao et al. [23] proposed the pyramid scene parsing network (PSPNet) for semantic segmentation. In order to incorporate suitable global features, they proposed pyramid pooling layer, which can aggregate the context information from different subregions and exploit the capability of global information. Pyramid pooling was commonly used in scene parsing task, while Park et al. [24] introduced it into EDSR for image SR task. They proposed EDSR-PP, an improved version of EDSR by applying four pyramid scales (1×1, 2×2, 3×3 and 4×4) into the upsampling module, and obtained better performance. Therefore, we adopt pyramid pooling layer to enhance the capability of exploiting the global context information, and to produce SR images with higher quality.

3 Our Method

Based on a state-of-the-art SR network, proSR [14], we propose a multi-scale progressive reconstruction network (MSPN) for image SR task. In this paper, we address some problems in proSR by several methods.

- We apply an improved inception module structure [25] into our proposed network and propose the multi-scale feature extraction block (MSFB). Unlike proSR using the improved dense compression units (DCU) to extract features, MSFB performs multiple filters with different kernel sizes on the same level for better image characterization.
- proSR removes batch normalization from DCU, causing the training process more difficult. However, in our proposed network, WN is applied into MSFB to address the gradient problem caused by simply stacking much layers. In addition, WN accelerates the convergence speed of training.
- Pyramid pooling layer is added before the convolution operation in the upsampling module. By aggregating multi-scale feature from different subregions, pyramid pooling layer improves the capability of obtaining global information. Furthermore, it enhances the utilization for local and global context information and achieves more competitive reconstruction performance.

3.1 Network Structure

As shown in Fig. 1, the overall architecture of MSPN is built as an asymmetric pyramid structure, which stacks more layers in low-level networks and fewer layers in high-layer networks. Our proposed network contains two parts: feature extraction block and image reconstruction block. We decompose the entire feature extraction process into a series of simpler pyramid modules $l_1, l_2, \ldots, l_s$ and each pyramid module performs a 2×2 upsampling of the input LR images. Let's denote I_{LR} as the input of MSPN, and $I_{out}^0, I_{out}^1, \ldots, I_{out}^s$ as the output of each pyramid module, where s denotes the scale factor. The mathematical formulation of each pyramid module l_i is:

$$I_{out}^i = f_{l_i}(I_{out}^{i-1}), \tag{1}$$

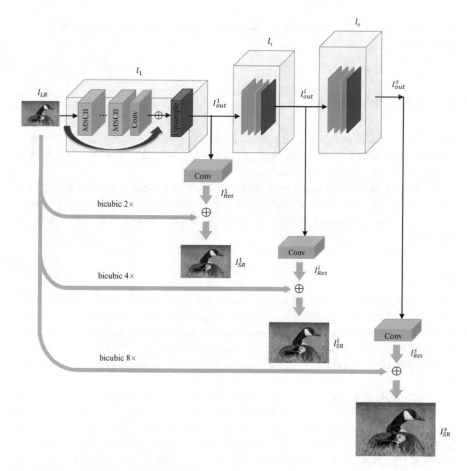

Fig. 1 The overall framework of MSPN

where f_{l_i} denotes the operations of the ith pyramid module. Specially, the input of the first pyramid module is I_{LR}.

After extracting hierarchical features with the pyramid modules, we stack a convolution layer to output the residual image I_{Res}, which can be obtained by

$$I_{Res}^i = f_{conv}(I_{out}^i), \qquad (2)$$

where f_{conv} denotes convolution operation.

At each pyramid module, the image reconstruction block delivers an upsampled image with different scale factors by residual skip connection to generate the output SR image I_{SR}:

$$I_{SR} = f_{up}(I_{LR}) + I_{Res}, \qquad (3)$$

where f_{up} denotes the operation of an upsampling kernel, here we use the bicubic operation.

3.2 Pyramid Module

The architecture of our proposed pyramid module is shown in Fig. 2. Our pyramid module consists of multi-scale compression blocks (MSCBs), global residual learning and the upsampling module.

Global residual learning is introduced to obtain feature maps before the upsampling module, and to further improve the information flows as there are several MSCBs in one pyramid module. The mathematical formation of local residual learning is:

$$F_{GF} = F_{LF} + I_{out}^{i-1}, \tag{4}$$

where I_{out}^{i-1} denotes the output of preceding pyramid module. In the i-th pyramid module, I_{out}^{i-1} is regarded as the initial input F_{CB}^0. After extracting low-level features F_{LF} in LR space, we produce the final feature maps F_{GF} through global residual learning. It should be noted that global residual learning can further encourage the flow of information and gradient, and help high-level networks acquire more effective features.

Multi-scale compression block (MSCB) is utilized to further exploit the hierarchical features and to enhance the discriminative representations. As shown in Fig. 3, it consists of multi-scale feature extraction blocks (MSFBs), compression layer and local dense feature fusion. MSFBs are designed for extracting multi-scale features, and MSCB fully utilizes these features to greatly improve the quality of SR image. However, directly using multi-scale features will cause enormous computational complexity. Therefore, we employ a 1×1 compression layer to adaptively fuse feature information from all the MSFBs, and to extract useful features before being passed to the next MSCB. Besides, we find that adding a compression layer before the image reconstruction module can improve the tightness of network.

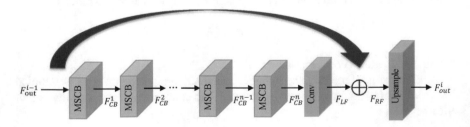

Fig. 2 The architecture of pyramid module

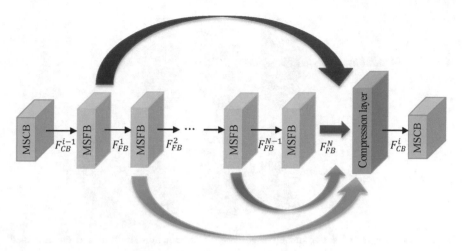

Fig. 3 Multi-scale compression block (MSCB)

Local dense feature fusion is applied to fuse features from all preceding MSFBs and to extract local and global features. After extracting hierarchical features in LR space, we stack a compression layer to achieve local dense feature fusion and to reduce the computational complexity. F_{CB}^{i-1} and F_{CB}^{i} represent the input and output of the ith MSCB respectively, and local dense feature fusion can be represented as

$$F_{CB}^{i} = C_{FF}(F_{CB}^{i-1}, F_{FB}^{0}, F_{FB}^{1}, \dots, F_{FB}^{N}), \tag{5}$$

where C_{FF} represents the function of the 1×1 convolution layer.

Upsampling module is utilized to obtain the feature maps with target scale factor in each pyramid module before being passed to the next pyramid module and the image reconstruction module. As the network depth grows, the receptive fields will also increase. However, the receptive fields are always not large enough to receive all the global context information during the training and testing process, resulting in poor performance.

Inspired by [23], we introduce pyramid pooling layer into our proposed upsampling module to incorporate the context information of different subregions and to improve the capability of exploiting global information. In Fig. 4, we respectively demonstrate the commonly used image reconstruction structure and our improved structure.

Due to its global contextual prior, pyramid pooling layer can greatly combine feature information with local and global context information to improve the image reconstruction performance. As shown in Fig. 5, pyramid pooling module proposed in [23] is designed with four different pyramid scales (1×1, 2×2, 3×3 and 6

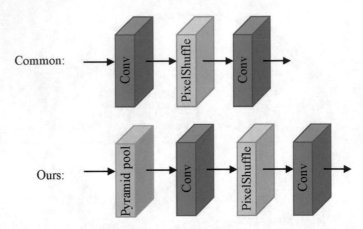

Fig. 4 Comparison of the commonly-used image reconstruction structure and our modified structure

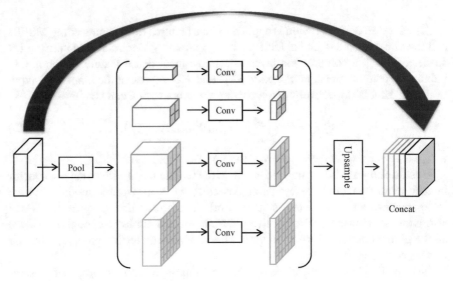

Fig. 5 The illustrations of pyramid pooling layer. In our proposed network, we remove the 6 × 6 convolution layer

× 6). However, we remove the 6 × 6 pyramid scale from pyramid pooling layer to reduce the computational complexity. In our proposed upsampling module, we perform multi-scale context information exploitation with three pyramid scales (1 × 1, 2 × 2 and 3 × 3).

3.3   *Multi-scale Feature Extraction Block*

The contents of different images differ greatly in size and position, and the color and structure in an image also vary considerably. Hence, it is difficult to select the appropriate kernel size for convolution layer. For image classification and detection task, Szegedy et al. [25] firstly proposed GoogLeNet with inception module. Different from previous works, it improves the performance by introducing a new network topology instead of stacking more layers.

As illustrated in Fig. 6, the inception module consists of three different filters and maximum pooling operation. Through different convolution layers, GoogLeNet can obtain distinctive feature information and improve the image characterization. Li et al. [26] adaptively introduced the inception module to extract different scale features for better image reconstruction performance. Furthermore, their method performed deeper and more effective exploitation for feature information in LR space by stacking the inception module.

Motivated by the method of applying inception module [25] into image SR network [26], we propose a dual-branch multi-scale feature extraction block (MSFB), as shown in Fig. 7. It forms various convolution kernel sizes by combining different kernels in two branches, and operates the feature maps on all the convolution layers. Besides, it achieves the feature sharing between two branches and concatenates all the outputs into deep feature maps. This approach can widen the network for better image characterization and feature utilization.

MSFB uses the dual-branch network to exploit the feature information by different convolution kernels. Let F_{FB}^{i-1} and F_{FB}^{i} be the input and output of the ith MSFB respectively. The mathematical formation of MSFB is:

$$A_1 = \sigma \left(F_{WN} \left(W_{3\times3} * F_{FB}^{i-1} \right) \right), \tag{6}$$

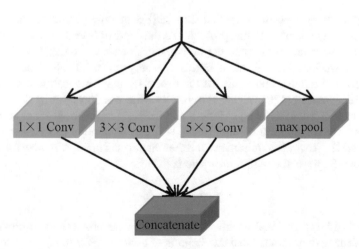

Fig. 6 The structure of Inception module

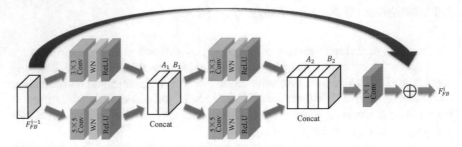

Fig. 7 Multi-scale feature extraction block (MSFB)

$$B_1 = \sigma\left(F_{WN}\left(W_{5\times5} * F_{FB}^{i-1}\right)\right), \tag{7}$$

$$A_2 = \sigma\left(F_{WN}(W_{3\times3} * [A_1, B_1])\right), \tag{8}$$

$$B_2 = \sigma\left(F_{WN}(W_{5\times5} * [A_1, B_1])\right), \tag{9}$$

$$F_{concat} = W_{1\times1} * [A_2, B_2], \tag{10}$$

where σ denotes the ReLU activation function. $W_{3\times3}$ and $W_{5\times5}$ are the weights of $W_{3\times3}$ and $W_{5\times5}$ convolution layers, where the bias term is omitted for simplicity.

$[\cdot, \cdot]$ denotes the concatenation operation. Besides, we apply the local residual learning to improve the efficiency of the network, which can be obtained by

$$F_{FB}^i = F_{FB}^{i-1} + F_{concat}. \tag{11}$$

Weight normalization is applied into MSFB module to address the gradient vanishing and gradient exploding problem. As the depth of neural networks increases, lack of normalization will make the network difficult to be trained. WN not only enables network to be trained at higher learning rates, but also speeds up the convergence of training and improves the results of image reconstruction.

Different from BN [20], WN [21] normalizes the weight vector by decomposing the weight vector into the weight length and the weight direction. Since it does not depend on the distribution of input data, WN can be applied to small mini-batch cases. Besides, it avoids requirement of different normalization formulas for training and testing. Suppose the output of image is

$$\mathbf{y} = \mathbf{w} \cdot \mathbf{x} + b, \tag{12}$$

where $\mathbf{w}$ is a k-dimensional weight vector, b is a scalar, and $\mathbf{x}$ is the k-dimensional vector of the feature input. Then the formula of applying WN to reparameterize the weight vector is

$$w = \frac{g}{v},$$ (13)

where v is a k-dimensional vector, g is a scalar, and v is the Euclidean norm of the vector v. In general, we assume that $v = g$, which is independent of the parameter v, so that the European norm of the weight vector can be fixed and the regularization effect will be performed. This weight decomposition method makes the training of network parameters more robust.

4 Experiment

In this section, we evaluate the performance of our proposed network MSPN on several benchmark datasets. The first subsection is to introduce the datasets used for training and testing. And in the second subsection, we introduce our implementation details and training details. Then, we analyze the structure of our proposed network from several aspects to show its performance. Finally, we give some quantitative and qualitative comparisons among our experimental results and other networks results.

4.1 Datasets

We use DIV2K dataset [15] as the training set, which contains 1000 h images that is divided into three parts: 800 training images, 100 validation images and 100 test images. For testing and benchmarking, we choose Set5 [28], Set14 [29], BSDS100 [30], Urban100 [31] and Manga109 [32] datasets, to verify the feasibility of our proposed network and to evaluate the performance of our image SR reconstruction results.

As most of the previous networks, all our training and testing are conducted on the luminance channel, and we choose the scale factors of 2 ×, 4 × and 8 × in this paper for training and testing.

4.2 Implementation and Training Details

For training our proposed network, we use RGB LR images whose patches are of size 64 × 64 as input. We randomly sample the LR patches and augment them in three ways: (1) random scaling the images with the ratio between 0.5 and 1.0; (2) random flipping the images horizontally or vertically; (3) random rotating with 90, 180 or 270°.

For 8 × scale factor, we build a three-stage progressive network and train the network with the curriculum-learning method. Each multi-scale compression block

(MSCB) consists of 8 multi-scale feature extraction blocks (MSFBs). We train the network with the mini batch-size of 8 and update it with Adam optimizer. In each training batch, 500 epochs are run with learning rates of 0.0001. We implement our network with Pytorch framework and use GTX 1080Ti to train the network.

4.3 Model Analysis

In Table 1, the comparison among the original network proSR, proSR with WN, proSR with pyramid pooling layer and our proposed MSPN is shown. It demonstrates how weight normalization and pyramid pooling layer can improve the network performance for $2 \times$, $4 \times$ and $8 \times$ scale factors.

(1) **Weight normalization**

We add weight normalization before each ReLU operation in multi-scale feature extraction block, which can achieve better performance than adding batch normalization or without using any normalization. Furthermore, weight normalization can provide more memory resources to our network for properly deepening network without additional computational complexity.

In Table 1, we can find that proSR with weight normalization gains better performance than the original proSR. This is mainly because the reparameterization of the weight vectors can alleviate the gradient problem and accelerate the optimization convergence.

Table 1 Comparison of proSR, with WN, with pyramid pooling layer (PPL), and MSPN for $8 \times$ scale factor (**bold** indicates the best performance)

Methods		Set5 PSNR/SSIM	Set14 PSNR/SSIM	BSDS100 PSNR/SSIM	Urban100 PSNR/SSIM
proSR	2×	37.95/0.9613	33.58/0.9126	32.10/0.9045	32.03/0.9341
With WN		38.02/0.9642	33.68/0.9133	32.17/0.9074	32.41/0.9355
With PPL		38.02/0.9655	33.68/0.9145	32.15/0.9056	32.45/0.9364
MSPN (our)		**38.06/0.9687**	**33.71/0.9264**	**32.21/0.9089**	**32.67/0.9388**
proSR	4×	32.28/0.9045	28.57/0.7813	27.55/0.7344	26.19/0.7961
With WN		32.39/0.9051	28.65/0.7846	27.60/0.7458	26.43/0.8022
With PPL		32.46/0.9067	28.67/0.7853	27.60/0.7463	26.48/0.8043
MSPN (our)		**32.50/0.9086**	**28.75/0.7902**	**27.64/0.7487**	**26.64/0.8077**
proSR	8×	27.16/0.7832	24.83/0.6454	24.76/0.5935	22.56/0.6287
With WN		27.29/0.7923	24.93/0.6589	24.80/0.6023	22.67/0.6323
With PPL		27.38/0.7948	24.91/0.6475	24.81/0.6098	22.71/0.6356
MSPN (our)		**27.45/0.7995**	**25.02/0.6523**	**24.86/0.6178**	**22.84/0.6432**

(2) **Pyramid pooling layer**

We apply pyramid pooling layer into the upsampling module to improve the visual effect of SR images. The poor quality of restored details in SR images always results from the absence of high-frequency information in LR images. The network cannot provide the receptive fields large enough to receive all the global context information. However, pyramid pooling layer can exploit more accurate high-frequency details by extracting multi-scale context information from different subregions.

In Table 1, we can validate that proSR with pyramid pooling layer yields better performance than the original proSR. And for some datasets, proSR with pyramid pooling layer also gains better results against proSR with WN. This is mainly because pyramid pooling layer can concatenate local and global context information to improve the accuracy of the recovered details and to produce SR images with higher quality.

4.4 Comparison with State-of-the-Art Methods

We compare our network with 11 state-of-the-art image super-resolution algorithms: Bicubic, A+ [33], SRCNN [5], VDSR [6], DRCN [7], DRRN [8], LapSRN [12, 13], EDSR [10], MSRN [26], CARN [34, 35], and proSR [14].

(1) **Quantitative Comparison**

The quantitative analysis is performed by comparing PSNR and SSIM in Table 2, which shows the results of our proposed network MSPN and other state-of-the-art algorithm for $2\times$, $4\times$ and $8\times$ scale factors. From this table, it can be found that MSPN performs better than most of state-of-the-art algorithms on all the testing datasets in each scale factors. This indicates that our network can achieve better reconstruction results by deeper information exploitation and more effective feature extraction. Besides, we have discussed that for the purpose of making full use of feature information in LR images, it is more effective to widen the network rather than deepen it. The observations illustrate that multi-scale feature extraction block and pyramid pooling layer allow our network to perform better features extraction, better image characterization and to produce more accurate reconstructed SR images.

(2) **Qualitative comparison**

In this section, we qualitatively compare the image reconstruction results among our network MSPN and other state-of-the-art algorithms. Figures 7 and 8 shows the visual comparisons for $4\times$ and $8\times$ image SR reconstruction results generated by our network and some other algorithms.

In Urban100 and BSDS100 datasets, most of the images contain various constructions such as high-rise buildings, subways, public facilities and infrastructures. The structures of these constructions are grid-like, or striped, or similar in color and shape, whose details are extremely difficult to recover. For image '253027' and

Table 2 Comparison between the result of our proposed MSPN and other methods (***bold italic*** indicates the best performance and **bold** indicates the second best performance.)

Methods	Scale	Set5 PSNR/SSIM	Set14 PSNR/SSIM	BSDS100 PSNR/SSIM	Urban100 PSNR/SSIM	Manga109 PSNR/SSIM
Bicubic	2×	33.68/0.9291	30.28/0.8684	29.58/0.8435	26.88/0.8439	30.83/0.9334
A+		36.58/0.9540	32.43/0.9060	31.23/0.8871	29.23/0.8955	35.41/0.9652
SRCNN		36.69/0.9553	32.39/0.9063	31.36/0.8881	29.51/0.8989	35.85/0.9676
VDSR		37.53/0.9589	33.05/0.9124	31.91/0.8966	30.77/0.9156	37.22/0.9730
DRCN		37.63/0.9586	33.07/0.9110	31.85/0.8947	30.74/0.9144	37.61/0.9218
DRRN		37.74/0.9593	33.23/0.9134	32.06/0.8971	31.23/0.9190	37.61/0.9734
LapSRN		37.52/0.9587	33.09/0.9128	31.40/0.8950	31.81/0.8952	37.27/0.9740
EDSR		*38.11/0.9601*	*33.92/0.9195*	*32.32/0.9012*	*32.33/0.9015*	*39.02/0.9767*
MSRN		38.06/0.9602	33.74/0.9167	32.24/0.9013	32.24/0.9014	38.82/0.9866
CARN		37.66/0.9585	33.48/0.9162	31.92/0.8961	30.98/0.9291	−/−
proSR		37.95/0.9655	33.58/0.9106	32.10/0.9097	32.03/0.9373	38.44/0.9865
MSPN(our)		**38.06/0.9689**	**33.75/0.9214**	**32.21/0.9126**	**32.67/0.9398**	**38.70/0.9883**
Bicubic	4×	28.41/0.8020	26.02/0.3943	25.95/0.6567	23.11/0.6602	24.87/0.7826
A+		30.32/0.8561	27.41/0.7445	26.80/0.7010	24.30/0.7205	27.02/0.8440
SRCNN		30.47/0.8568	27.55/0.7443	26.93/0.6994	24.55/0.7234	27.60/0.8545
VDSR		31.33/0.8790	28.04/0.7618	27.34/0.7166	25.17/0.7532	28.83/0.8810
DRCN		31.52/0.8806	28.10/0.7620	27.21/0.7151	25.17/0.7540	28.86/0.8806
DRRN		31.55/0.8877	28.19/0.7714	27.33/0.7278	25.44/0.7627	29.13/0.8920
LapSRN		31.53/0.8810	28.11/0.7628	27.32/0.7280	25.21/0.7550	29.02/0.8840
EDSR		**32.42/0.8958**	**28.74/0.7866**	*27.73/0.7420*	**26.60/0.8033**	*31.05/0.9438*
MSRN		32.04/0.8896	28.50/0.7749	27.50/0.7271	26.02/0.7889	30.17/0.9033
CARN		32.03/0.8924	28.52/0.7760	27.47/0.7323	25.88/0.7788	−/−
proSR		32.38/0.9075	28.57/0.7898	27.55/0.7334	26.19/0.7945	30.58/0.9146
MSPN (our)		*32.50/0.9143*	*28.76/0.7904*	**27.64/0.7478**	*26.64/0.8099*	**30.95/0.9187**
Bicubic	8×	24.40/0.6485	23.10/0.5663	23.66/0.5482	20.72/0.5164	21.45/0.6135
A+		25.53/0.6648	23.98/0.5530	24.21/0.5155	21.33/0.5190	22.37/0.6450
SRCNN		25.32/0.6571	23.76/0.5914	24.13/0.5665	21.29/0.5438	22.44/0.6784
VDSR		25.93/0.7243	24.24/0.6172	24.50/0.5832	21.66/0.5704	22.83/0.6763
DRCN		25.73/0.6743	24.21/0.5515	24.48/0.5170	21.66/0.5288	23.30/0.6687
LapSRN		26.15/0.7382	24.35/0.6210	24.51/0.5864	21.77/0.5811	23.41/0.7071
EDSR		27.09/0.7810	**24.94/0.6431**	**24.81/0.5983**	22.50/0.6237	**24.70/0.7834**
MSRN		26.59/0.7253	24.90/0.5962	24.72/0.5401	22.34/0.5967	24.27/0.7510
CARN		26.74/0.7667	24.83/0.6356	24.73/0.5911	22.25/0.6044	−/−
proSR		**27.16/0.7832**	24.83/0.6454	24.76/0.5935	**22.56/0.6287**	24.67/0.7832
MSPN (our)		*27.45/0.7995*	*25.02/0.6523*	*24.86/0.6178*	*22.84/0.6432*	*24.98/0.7945*

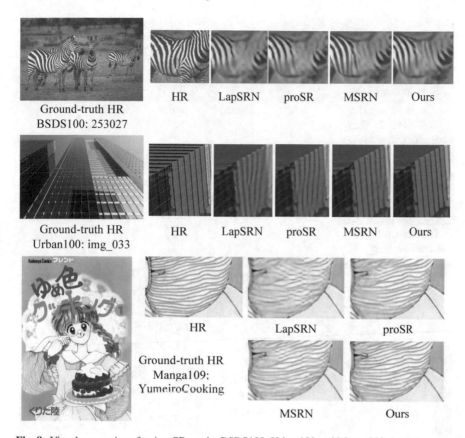

Fig. 8 Visual comparison for 4 × SR on the BSDS100, Urban100 and Manga109 datasets

image 'img_033', other algorithms usually produce blurring artifacts and twisted lines in images, especially for the striped or grid-like contents.

What's worse, the generated lines are completely inconsistent with the direction of the original lines. However, MSPN can solve these problems and recover more real details and distinct features.

Besides, for image 'TotteokiNoABC' in Manga109 dataset, we can find that some methods suffer from blurring artifacts and over-smoothed edges around the letters. By contrast, MSPN shows great abilities in recovering more accurate information and clear outline, which are closer to the real images (Fig. 9).

In general, our proposed method demonstrates better performance both in quantitatively and visually, whether for 4× or 8× scale factors. Our improved feature extraction module greatly achieves deeper and more effective information exploitation in LR images, and the image reconstruction module makes full use of feature information and spatial relationship to reconstruct high-quality SR images, which is more faithful to the real images.

194   S. Ying et al.

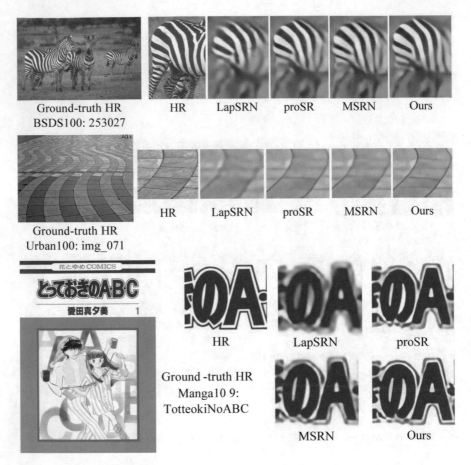

Fig. 9 Visual comparison for 8 × SR on the BSDS100, Urban100 and Manga109 datasets

5   Conclusion

In this paper, we propose a multi-scale progressive reconstruction method of image
SR. Aiming at scarcity of the high-frequency details in single LR image for different
scale factors, our proposed network is designed as the asymmetric pyramid structure
to extract abundant low-level feature information and to reconstruct high-quality
SR images step by step. The multi-scale feature extraction block is used to widen
the network for deeper and more effective information exploitation. Besides, weight
normalization is applied to address the gradient vanishing and gradient exploding
problem, and to accelerate the convergence speed of training. Moreover, pyramid
pooling layer serves to improve the accuracy of the image reconstruction results
by concatenating local and global context information. Extensive evaluations on
benchmark datasets have shown that our proposed algorithm gains great performance
against the state-of-the-art algorithms in image SR.

References

1. Shi WZ, Caballero J, Ledig C, Zhuang XH, Bai WJ, Bhatia K, de Marvao AMSM, Dawes T, O'Regan D, Rueckert D (2013) Cardiac image super-resolution with global correspondence using multi-atlas patchmatch. Proceedings of MICCAI, pp 9–16
2. Gunturk BK, Altunbasak Y, Mersereau RM (2004) Super-resolution reconstruction of compressed video using transform-domain statistics. IEEE Trans Image Process 13:33–43
3. Zou WWW, Yuen PC (2012) Very low resolution face recognition problem. IEEE Trans Image Process 21:327–340
4. Yıldırım D, Güngör O (2012) A novel image fusion method using ikonos satellite images. J Geodesy Geoinform 427–429:1593–1596
5. Dong C, Loy CC, He K, Tang X (2014) Learning a deep convolutional network for image super-resolution. Eur Conf Comput Vis 8692:184–199
6. Kim J, Kwon Lee J, Mu Lee K (2016) Accurate image super-resolution using very deep convolutional networks. In: Proceedings of the IEEE conference on computer vision and pattern recognition, pp 1646–1654
7. Kim J, Kwon Lee J, Mu Lee K (2016) Deeply-recursive convolutional network for image super-resolution. In: Proceedings of the IEEE conference on computer vision and pattern recognition, pp 1637–1645
8. Tai Y, Yang J, Liu X (2017) Image super-resolution via deep recursive residual network. In: Proceedings of the IEEE conference on computer vision and pattern recognition, pp 2790–2798
9. Tong T, Li G, Liu XJ, Gao QQ (2017) Image super-resolution using dense skip connections. In: Proceedings of the IEEE international conference on computer vision, IEEE computer society, pp 4809–4817
10. Lim B, Son S, Kim H, Nah S, Lee KM (2017) Enhanced deep residual networks for single image super-resolution. In: The IEEE conference on computer vision and pattern recognition workshops, vol 1, pp 1132–1140
11. Zhang YL, Tian YP, Kong Y, Zhong BN, Fu Y (2018) Residual dense network for image super-resolution. In: The IEEE/CVF conference on computer vision and pattern recognition, pp 2472–2481
12. Lai WS, Huang JB, Ahuja N, Yang MH (2017) Deep Laplacian pyramid networks for fast and accurate super-resolution. In: IEEE conference on computer vision and pattern recognition, pp 5835–5843
13. Lai WS, Huang JB, Ahuja N, Yang MH (2018) Fast and accurate image super-resolution with deep Laplacian pyramid networks. IEEE Trans Pattern Anal Mach Intell 99
14. Wang YF, Perazzi F, Mcwilliams B, et al (2018) A fully progressive approach to single-image super-resolution. In: Proceedings of the IEEE/CVF conference on computer vision and pattern recognition workshops, pp 977–986
15. Agustsson E, Timofte R (2017) Ntire 2017 challenge on single image super-resolution: dataset and study. In: The IEEE conference on computer vision and pattern recognition workshops, vol 3, p 2
16. Keys RG (1981) Cubic convolution interpolation for digital image processing. IEEE Trans Acoust Speech Signal Process 37:1153–1160
17. Shi W, Caballero J, Husz´ar F, Totz J, Aitken AP, Bishop R, Rueckert D, Wang Z (2016) Real-time single image and video super-resolution using an efficient sub-pixel convolutional neural network. In: Proceedings of the IEEE conference on computer vision and pattern recognition, pp 1874–1883
18. Dong C, Loy CC, Tang X (2016) Accelerating the super-resolution convolutional neural network. In: European conference on computer vision. Springer, pp 391–407
19. Ledig C, Theis L, Husz´ar F, Caballero J, Cunningham A, Acosta A, Aitken A, Tejani A, Totz J, Wang Z, et al (2017) Photo-realistic single image super-resolution using a generative adversarial network. In: The IEEE conference on computer vision and pattern recognition, pp 105–114

20. Sergey I, Christian S (2015) Batch normalization: Accelerating deep network training by reducing internal covariate shift. In: International conference on machine learning, pp 448–456
21. Tim S, Diederik PK (2016) Weight normalization: A simple reparameterization to accelerate training of deep neural networks. Adv Neural Inform Process Syst 901–909
22. Yu JH, Fan YC, Yang JC et al (2018) Wide activation for efficient and accurate image super-resolution. In: IEEE conference on computer vision and pattern recognition
23. Zhao H, Shi J, Qi X, Wang X, Jia J (2017) Pyramid scene parsing network. In: The IEEE conference on computer vision and pattern recognition, pp 2881–2890
24. Park D, Kim K, Chun SY (2018) Efficient module based single image super resolution for multiple problems. Proceedings of CVPRW, pp 995–1003
25. Szegedy C, Liu W, Jia Y, Sermanet P, Reed S, Anguelov D, Erhan D, Vanhoucke V, Rabinovich A (2015) Going deeper with convolutions. In: Proceedings of the IEEE conference on computer vision and pattern recognition, pp 1–9
26. Li JC, Fang FM, Mei KF, Zhang GX (2018) Multi-scale residual network for image super-resolution. In: European conference on computer vision. Springer, pp 527–542
27. Huang G, Liu Z, Weinberger KQ, van der Maaten L (2017) Densely connected convolutional networks. Proceedings of CVPR, pp 2261–2269
28. Bevilacqua M, Roumy A, Guillemot C, Alberi-Morel ML (2012) Low-complexity single-image super-resolution based on nonnegative neighbor embedding. In: Proceedings of the 23rd British machine vision conference
29. Zeyde R, Elad M, Protter M (2010) On single image scale-up using sparse representations. In: International conference on curves and surfaces. Springer, pp 711–730
30. Arbelaez P, Maire M, Fowlkes C, Malik J (2011) Contour detection and hierarchical image segmentation. IEEE Trans Pattern Anal Mach Intell 33:898–916
31. Huang JB, Singh A, Ahuja N (2015) Single image super-resolution from transformed self-exemplars. In: Proceedings of the IEEE conference on computer vision and pattern recognition, pp 5197–5206
32. Matsui Y, Ito K, Aramaki Y, Fujimoto A, Ogawa T, Yamasaki T, Aizawa K (2017) Sketch-based manga retrieval using manga109 dataset. Multimedia Tools Appl 76:21811–21838
33. Timofte R, De Smet V, Van Gool L (2014) A+: adjusted anchored neighborhood regression for fast super-resolution. In: Asian conference on computer vision. Springer, pp 111–126
34. Ahn N, Kang B, Sohn KA (2018) Fast, accurate, and lightweight super-resolution with cascading residual network. In: European conference on computer vision. Springer, pp 256–272
35. Ahn N, Kang B, Sohn KA (2018) Image super-resolution via progressive cascading residual network. In: 2018 IEEE/CVF conference on computer vision and pattern recognition workshops, pp 904–912

Track Related Bursty Topics in Weibo

Yuecheng Yu, Yu Gu, Ying Cai, Daoyue Jing, and Dongsheng Wang

Abstract Weibo has become an important means for people to share, disseminate and obtain information in real time. BTM can effectively discover bursty topics in Weibo, but cannot track the related bursty topics. Based on the time series of Weibo, the binary word pair is used for topic modeling and the bursty topics in the Weibo are extracted to form a new topic set. And then, the similarity of the topics in adjacent time periods can be calculated by KL metric and the related bursty topics can be tracked. The experimental results show that the method can effectively segment the time series of Weibo topics, and realize the discovery and tracking of related topics in Weibo.

Keywords Bursty topics · BTM model · Topic tracking · Topic tracking

1 Introduction

Social media has become the main way of information acquisition and information dissemination in the past decade [1]. A variety of social media platforms provide convenience for people to share, communicate, and collaborate information [2]. As a social networking platform, Wcibo gradually become the preferred media for people to express their opinions and report unexpected incidents [3]. Since the spread of information in Weibo is very fast, many sudden topics are often first revealed in Weibo [4]. Then the detection and discovery of sudden topics in Weibo has become a new research hotspot [5].

It should be noted that the detected bursty topic in Weibo tends to evolve over time. Thus, it is not only necessary to discover sudden topics in time, but also to effectively evaluate the evolution of topics in Weibo. The traditional topic evolution method is mainly used for the processing of long texts with clear contexts such as news and corpus [6–9]. Weibo content is also composed of a series of texts, text-based topic

Y. Yu (✉) · Y. Gu · Y. Cai · D. Jing · D. Wang
School of Computer Science, Jiangsu University of Science and Technology, Zhenjiang, China
e-mail: zhjyuyuecheng@163.com

© Springer Nature Switzerland AG 2021
H. Lu (ed.), *Artificial Intelligence and Robotics*,
Studies in Computational Intelligence 917,
https://doi.org/10.1007/978-3-030-56178-9_15

discovery technology is still the mainstream technology for microblogging topics [10–12], some of them are various improved versions based on LDA [13–15].

In the process of processing Weibo, these methods still capture the word co-occurrence patterns from the document level, which will face serious data sparsity when dealing with microblog short text [15, 16]. In this paper, we will expand the short text topic modeling method proposed in the BTM [16] to track the related bursty topics and its evolution process.

2 Related Work

In the BTM model, the topic is modeled using a binary phrase as the basic unit [16]. In the Weibo scenario, let a short text set $B = \{b_1, ..., b_{NB}\}$, assume that it contains N_B biterm words and K topics over W unique words in the collection. Let $b_i = (w_{i,1}, w_{i,2})$ denote biterm word pair, $z \in [1, K]$ be a topic indicator variable which represents the biterm word pairs correspond to the topic. The word distribution for topics is φ, which is a $K \times W$ matrix, and φ_k denotes the kth row of the matrix. In fact, φ_k is a W-dimensional multinomial distribution. θ is a K-dimensional multinomial distribution and represent the prevalence of topics in the collection. Therefore, by using the Collapsed Gibbs algorithm [17] to estimate the model parameters, the discovery of the bursty topic in Weibo is realized.

3 Tracking Method of Weibo Bursty Topic Evolution

Assume that n_b^t is the number of times that the biterm word pair b appears in the published microblog in the time slice t, and $\overline{n}_b^t$ is the average of n_b^t, that is $\overline{n}_b^t = \frac{1}{S}\sum_{s=1}^{S} n_b^{t-s}$. Let η_b^t be the probability that the topic of the biterm word pair b is the same as the topic of the bursty topic, then it can be estimated according to formula (1), where ε is a relatively small positive number used to avoid the probability of a zero value.

$$\eta_b^t = \frac{\max[(n_b^t - \overline{n}_b^t), \varepsilon]}{n_b^t} \tag{1}$$

Specifically, let h_z denote a topic selector, which is used to indicate whether the topic is the same as the detected bursty topic. In our method, the value of h_z is sampled from the Bernoulli distribution. The topics are defined as related topics, and represented by the data set $E_z = \{z \mid h_z = 1, z = 1, ..., K\}$.

To illustrate whether the topic of the biterm word pair is the related topic, a binary indicator variable Y_i is introduced. If $Y_i = 1$, then the topic of the word pair is the related topic. Otherwise, the topic of the word pair is not the related topic.

The algorithm of our method can be described as follows:

(1) For each biterm word $b_i \in B$ of each document, the generated document (biterm set) can be obtained by sampling $\eta \sim$ Beta (γ_0, γ_1) and topic selector $h_z \sim$ Bern (η), where $\overrightarrow{h} = \{h_z\}_{k=1}^{K}$, then according to $\theta_i \sim$ Dir (α), the parameter of document-topic distribution θ_i can be obtained.

(2) For each topic $k \in \{1, 2, ..., K\}$, if the topic is not a related topic, sampling and generating the background word distribution parameter $\varphi_0 \sim$ Dir (β), otherwise, sampling and generating topic "entry" Multiple distribution parameter $\varphi_k \sim$ Dir (β).

(3) For each biterm pair $<w_{i,1}, w_{i,2}>$ of each word pair $b_i \in B$, sampling and generating category label $Y_i \sim$ Bern (η_{b_i}):

    (a) If $Y_i = 0$, the sample generates a "term" biterm pair: $w_{i,1} \sim$ Multi (φ_0), $w_{i,2} \sim$ Multi (φ_0);

    (b) If $Y_i = 1$, sample a related topic $z \sim$ Multi (θ), sample words pairs $w_{i,1} \sim$ Multi (φ_z), $w_{i,2} \sim$ Multi (φ_z).

Similar to BTM, model parameters are still estimated using the Collapsed Gibbs algorithm [17]. In order to track the evolution of bursty topics, bursty topics that has been detected and the topic captured by the evolution tracking model are combined into a new topic set, and then the Weibo content in the topic set is re-segmented by time slice. Lets $sub_1 = \{w_{1,1}, ..., w_{1,n}\}$ and $sub_2 = \{w_{2,1}, ..., w_{2,n}\}$ be subtopics in two adjacent time slices, and $P(i)$ represent the probability distribution of the i-th word in subtopic sub_1, $Q(i)$ is the probability distribution of the i-th word in subtopic sub_2, then the KL distances between the two topics can be calculated according to formula (2) and newly evolved topic can be dectected [18].

$$D(P\|Q) = \sum_i \ln(\frac{P(i)}{Q(i)})P(i) \qquad (2)$$

4 Experimental Analysis

In order to verify the effectiveness of the proposed method in tracking the evolution of microblogging topic, we use the relevant corpus of the microblogging topic "#Chongqing Wanzhou Bus Falling River#" as the experimental data set. Table 1 describes the related topics in each time slice, and fully describes the entire evolution of the topic.

Figure 1 is the KL distance between different topics in adjacent time slices. As shown in Fig. 1, the KL distance between adjacent time slices visually reflects the evolution and transition between different topics in Table 1.

When the bursty topic "Chongqing Wanzhou bus crashed into the river" just appeared, the focus of the topic was mainly on the visual description of the bus crash accident. When the topic progressed to the second time slice, Weibo users began to

Table 1 Specifies the evolutionary tracking of bursty topics

Time slice	Evolution tracking of bursty topics	Topic descriptors
1	# Chongqing Wanzhou bus crashes into the river#	Chongqing, bus, Wanzhou, traffic accident, broken, guardrail, Yangtze River, frontal crash, car
2	# Chongqing bus female driver#	Car, female driver, Vehicle owner, collision, driving age, rumors, public opinion, misunderstanding, irresponsibility
3	# Falling River Bus Found#	Location of the river, underwater, determined, length 11 meters, width 3 meters, 22 roads, positioning, salvage
4	# Video of bus crashing into the river frontal impact#	Recorder, impact, collision, front, instant, rushing, guardrail, Yangtze River
5	#Bus Responsibility for Falling Down the River#	Responsibility, credit card, boarding, confirmation, record, responsibility, driver, salvage, number of passengers
6	#15 people lost their connection in Chongqing River Falling Accident#	Loss of connection, rescue, readiness, preparation, 800 tons, hoisting, lifting, salvaging
7	# Falling river bus black box#	Black box, salvage, confirmation, exposure, video, interior, condition, steering wheel, picture
8	#Reasons for Chongqing Bus Falling Down the River#	Passengers, drivers, fierce, disputes, mutual control, loss of control, quarrels, monitoring, display
9	#Female Bus Passengers in Chongqing#	Passing stops, female passengers, beatings, assaults, excesses, costs, uncontrollability, crashes

pay attention to the new focus. As shown in Fig. 1, since the differences in topics in the two time slices are large, the corresponding KL values in the two time slices also differ greatly. In the second and third time slice, although the topics discussed were different, it was mainly around the bus crash. From time slice 3–6, the KL value varies greatly. In fact, the focus of the discussion was also from the mistaken belief that the car retrograde caused the bus to fall into the river and gradually turned into a car driver. When the content of the Weibo topic changes smoothly, and the corresponding KL values are relatively close.

5   Conclusion

Based on the traditional text probability model BTM, the method of tracking related bursty topics in Weibo is proposed in this paper. With the help of the relevance of Weibo content in time, the sudden topics discovered by BTM model are recombined

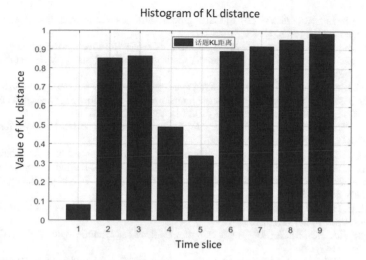

Fig. 1 KL distance histogram

into a new topic set. Then, by calculating the similarity of the burst topics on the adjacent time slices, the effective tracking of the related bursty topics is realized. This method is helpful to observe the hot topics that Weibo users pay attention to and predict the trend of public opinion in real time. This also provides effective help for realizing the effective guidance of public opinion and tracing the sequence of hot events.

Acknowledgements This work is supported by National Natural Science Foundation of China No. 61702234, Science and Technology Support Project (Social Development) in Jiangsu Province No. BE2014692. Science and Technology Support plan (Social Development) in Zhenjiang City No. SH2015018.

References

1. Stieglitz S, Mirbabaie M, Ross B et al (2018) Social media analytics—challenges in topic discovery, data collection, and data preparation. Int J Inf Manage 39:156–168
2. Yu R, Qiu H, Wen Z et al (2016) A survey on social media anomaly detection. ACM SIGKDD Explor Newsl 18(1):1–14
3. Fan R, Zhao J, Xu K (2015) Topic dynamics in Weibo: a comprehensive study. Soc Netw Anal Min 5(1):41 (2015)
4. Kaplan AM, Haenlein M (2010) Users of the world, unite! the challenges and opportunities of social media. Bus Horiz 53(1):59–68
5. Lin CX, Zhao B, Mei Q et al (2010) PET: a statistical model for popular events tracking in social communities. In: Proceedings of the 16th ACM SIGKDD international conference on knowledge discovery and data mining, ACM, pp. 929–938
6. Xujian Z, Chunming Y, Bo L (2014) a news topology evolution mining method based on feature evolution. Chin J Comput 4:819–832

7. Jensen S, Liu XZ, Yu YG (2016) Generation of topic evolution trees from heterogeneous bibliographic networks. J Inf 4(2):606–621
8. Jo Y, Hopcroft JE, Lagoze C (2011) The web of topics: discovering the topology of topic evolution in a corpus. WWW 2011-session: spatio-temporal analysis. ACM, Hyderabad, India, pp. 257–266 (2011)
9. Lin C, Lin C, Li J et al (2012) Generating event storylines from microblogs. In: Proceedings of the 21st ACM international conference on information and knowledge management. ACM, New York, pp 175–184. https://doi.org/10.1145/2396761.2396787
10. Wei X, Bin Z, Genlin J (2016) Microblog topic evolution algorithm based on forwarding relationship. Comput Sci 43(2):79–100
11. Mei Q, Zhai CX (2005) Discovering evolutionary theme patterns from text: An exploration of temporal text mining. In: Proceedings of the eleventh ACM SIGK-DD international conference on knowledge discovery in data mining. ACM, New York, pp 198–207. https://doi.org/10.1145/1081870.1081895
12. Yanli H, Liang Z, Weiming Z (2012) A method of modeling and analysis of topic evolution. Acta Autom Sinica 38(10):1690–1697
13. Ying F, Heyan H, Xin X (2014) topic evolution analysis for dynamic topic numbers. Chin J Inf Sci 28(3):142–149
14. Jayashri M, Chitra P (2012) Topic clustering and topic evolution based on temporal parameters. In: International conference on recent trends in information technology, IEEE, Chennai, India, pp 559–564 (2012)
15. Blei DM, Ng AY, Jordan MI et al (2003) Latent dirichlet allocation. J Mach Learn Res 3:2003
16. Yan X, Guo J, Lan Y, Cheng X-Q (2013) A biterm topic model for short texts. In: Proceedings of the 22nd international conference on world wide web. Rio de Janeiro, Brazil, pp 1445–1456 (2013)
17. Zaho B, Xu W, Ji GL (2016) Discovering topic evolution topology in a microblog corpus. In: Third international conference on advanced cloud and big data. CBD, China, pp 7–14 (2016)
18. Griffiths TL, Steyvers M (2004) Finding scientific topics. Natl Acad Sci 101(1):5228–5235

Terrain Classification Algorithm for Lunar Rover Based on Visual Convolutional Neural Network

Lanfeng Zhou and Ziwei Liu

Abstract With the development of society, the lunar exploration capability marks the development of aerospace science and technology, and it has received more and more attention. We propose a method which combines vision-based convolutional neural networks with ensemble learning to classify the current terrain by the pictures taken by the onboard camera of the lunar rover. According to the classification result, the lunar rover can independently select a better path, avoiding the unnecessary trouble caused by the delay of the land-month communication. The overall accuracy of our classification is 80%, and some of them have higher precision. It is expected that the classification results will help the decision of path planning.

Keywords Terrain classification · Convolutional neural network · Vision-based · Ensemble learning

1 Introduction

Terrain classification is important in the driving process of a lunar rover, especially in complex terrain environments. If the potential slip corresponding to the type of terrain is not taken into account when evaluating the path or when planning the path, it will have a great impact on the path planning.

The lunar rover mainly uses teleoperation and autonomous operation when performing tasks. For the teleoperation mode, the lunar rover will send the surrounding terrain information to the ground teleoperation platform. We can monitor and control the state of the lunar rover with the assistance of the ground staff. We plan the next behavior of the parade vehicle, and send it in the form of instructions. For the lunar rover, which is performing the task, performs the next action

L. Zhou (✉) · Z. Liu
Shanghai Institute of Technology, Shanghai, China
e-mail: lfzhou@sit.edu.cn

Z. Liu
e-mail: lzwsafm@gmail.com

© Springer Nature Switzerland AG 2021
H. Lu (ed.), *Artificial Intelligence and Robotics*,
Studies in Computational Intelligence 917,
https://doi.org/10.1007/978-3-030-56178-9_16

by the obtained teleoperation command. Although, the teleoperation reduces the performance requirements of the lunar rover. But, the teleoperation has obvious defects, such as the communication delay is large. The two-way delay of the lunar communication is at least 2–3 s. Considering the time lag of data processing, the communication delay can reach 15–20 s. In this way, communication is inefficient and communication time is limited.

On the other hand, the lunar rover needs to be able to adapt to the complex lunar surface and carry scientific instruments to perform tasks such as detection, sampling, and carrying. Under such requirements, it is difficult to complete complex tasks by means of teleoperation. This requires a cruiser with a limited life span to have a greater autonomous operation capability to improve the autonomous performance of the lunar rover and expand the detection range of the lunar rover.

The premise of the lunar rover's ability to improve its autonomous operation is to effectively obtain information about the surrounding obstacles. A lunar rover with autonomous operation capability should be able to analyze the environment information surrounding it obtained by on-board cameras and other sensors under the condition of clear target points, identify possible obstacles, estimate the traversability of surrounding terrain, and plan a way to avoid obstacles. A reliable path to the target point, and more importantly, it can stop, observe the surrounding environment, re-plan the path according to changes in the environment, and do not need too much assistance from the Earth.

We have reviewed a lot of literature [1–5] to solve this problem. In this paper, we proposes a vision-based convolutional neural network terrain classification method, which will enable the robot system to learn the type of terrain in the field of view through training and learning. We first cleaned and mapped the terrain camera images of the lunar rover from the Chang'e III for model training. After the training is completed, the lunar rover can automatically extract the picture features through the convolutional neural network and make inferences to distinguish the terrain categories.

In Sect. 2, we studied previous methods of terrain classification and the theoretical basis of CNN, which forms the basis of this study. In Sect. 3, we describe the monthly terrain classification algorithm proposed in this paper. In Sect. 4, we describe the experiments and results. Section 5, we presents the field of future research and summarizes the paper.

2 Related Work

2.1 Image Classification

In recent years, automatic classification techniques based on neural networks have been fully developed. From 2012, on the competition of ImageNet [6], AlexNet [7] was proposed. On the ICLR2015, the VGG [8] was proposed. On the CVPR2018,

Google proposed NasNet [9], it was training on 500 GPUs. The accuracy of top5 and top1 of the competition of ImageNet increased from 57.1 and 76.3% to 96.2 and 82.7%. It can be seen that the accuracy of image classification based on convolutional neural networks is constantly improving.

2.2 Terrain Classification

MIT's Karl Iagnemma classifies the terrain into three categories [10], rock, sand and beach grass, which gave good results. They had higher accuracy in sand and beach grass, but lower accuracy in distinguishing rocks.

Lauro Ojeda of the University of Michigan also did some research on terrain classification and some terrain description [11]. They used a fully connected neural network with only one hidden layer to classify the terrain. They divided into five categories. They are gravel, grass, sand and pavement dirt. The final average highest accuracy rate reached 78.4%.

2.3 Convolutional Neural Network

Neural networks are proposed by human analog neurons to transmit signals [12, 13]. Convolutional neural networks are a special structure of neural networks, which are widely used in the field of imaging. The image is divided into small regions in the same manner as the brain perceives the object, and the features of each region are learned to classify the input image [14, 15]. CNN is mainly composed of a convolution layer, a pooling layer and a fully connected layer [16].

A simple neural network, a chained structure in which each layer is a function of the previous layer [17, 18]. The first layer:

$$H^{(1)} = g^{(1)}(W^{(1)T}x + b^{(1)})$$

$H^{(1)}$ is the output of the first layer, $g^{(1)}$ is a nonlinear variation function, $W^{(1)}$ is the weight matrix, which is the value we need to train. b is bias, it is also the model that needs to be trained. x is an input layer, it is a vector. The second layer:

$$H^{(2)} = g^{(2)}(W^{(2)T}H^{(1)} + b^{(2)})$$

$H^{(1)}$ is the output of the first layer. So, the nth layer of network output we can express as:

$$H^{(n)} = g^{(n)}(W^{(n)T}H^{(n-1)} + b^{(n)})$$

Fig. 1 Soft gravel

There is a theory that when you can represent an arbitrary function by a neural network which has more than 2 layers. But from a practical point of view, training a deep network requires much less parameter than training a shallow network. The reason is that the shallower network has already extracted the basic features, and the higher layer network only needs to combine these basic features to get more complex features. It is similar to modularization in industrial production [19, 20].

3 Lunar Terrain Classification

3.1 Data Processing and Labeling

We obtained the monthly data of the Chang'e III from the network. After cleaning, cutting and labeling, we got 4,801 which are size of 784 × 576 × 3 valid sample image with label. According to the slipping effect of the lunar rover in different situations, the 4,801 sample images were divided into four categories, soft gravel topography, compacted soil topography, rocky terrain and concave land. Among them, the compacted soil topography is the optimal travel choice, and the soft gravel topography has a large slip ratio. Rocky terrain and concave terrain are terrains that are to be avoided as much as possible. The following four sets of figures show four types of sample features (Figs. 1, 2, 3 and 4).

The data type and quantity are as follows (Table 1).

3.2 Model

We used the classic AlexNet as a basis for improvement [7]. We divided the samples into three groups and used three AlexNet of the same structure to extract and classify each feature. It is equivalent to fitting a classification standard to each model, and preliminary classification of the data to obtain its own classification results. In the idea of using integrated learning, combined with the preliminary structure of the

Fig. 2 Compacted soil

Fig. 3 Rocky terrain

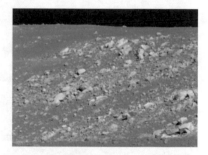

Fig. 4 Concave land

Table 1 The data type and quantity of our dataset

Class	Soft gravel	Compacted soil	Rocky terrain	Concave land
Number	1043	2765	625	368

three models, vote. The voting weight of each model result is non-linear. We fit with a simple fully connected neural network. The model is as follows (Fig. 5).

The following is a detailed description of the model:

The first layer is an input layer. It is a picture, which size is $784 \times 576 \times 3$.

The second layer is a convolutional layer. The input layer is convoluted using 96 convolution kernels which size is $11 \times 11 \times 3$ with a step size of 4 and 96 bias. The

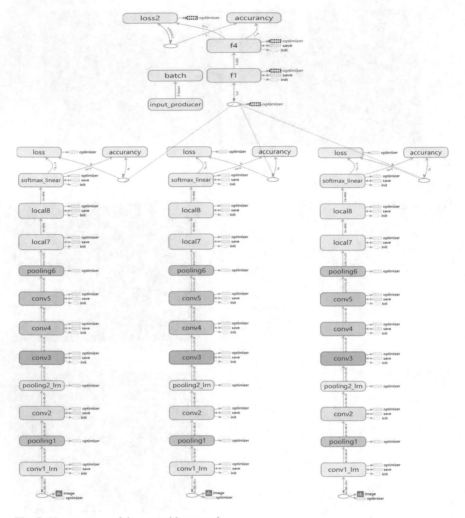

Fig. 5 The structure of the ensemble network

convolution formula is as follows:

$$p_{i,j} = f\left(\sum_{d=0}^{D-1}\sum_{m=0}^{F-1}\sum_{n=0}^{F-1} w_{d,m,n} x_{d,i+m,j+n} + w_b\right)$$

$p_{i,j}$ is the pixel value of row i and column j; D is the depth of the convolution kernel; F is the size of the convolution kernel; $w_{d,m,n}$ is the weight of the convolution kernel in m rows and n columns; w_b is the bias term. Use the ReLU activation function after the convolution calculation is complete:

$$f(x) = \max(0, x)$$

x is the input parameter; after the activation result is obtained, the Local Response Normalization operation is performed:

$$p^i_{x,y} = a^i_{x,y} / (k + \alpha \sum_{j=\max(0,i-n/2)}^{\min(N-1,i+n/2)} (a^j_{x,y})^2)^\beta$$

$a^i_{x,y}$ denotes the output of the i-th convolution kernel calculated by the ReLU activation function at position x row y column. n is the number of adjacent kernel maps at the same position. N is the number of convolution kernels; $k = 2, \alpha = 10^{-4}, n = 5, and \beta = 0.75$. After obtaining the convolution calculation result, the maximum pooling calculation is performed on the result:

$$H = D^{\lambda,\tau}_{\max}(C) = \text{maxdown}_{\lambda,\tau}(C)$$

C is the input convolution area, and the block size is $\lambda \times \tau$ not repeated sampling. Find the maximum value in this area to represent this area. $\lambda = \tau = 3$ and the step size is 2. Get the most pooled results.

The third layer is a convolutional layer. The results of the second layer are convoluted using 384 which size is $3 \times 3 \times 256$ convolution kernels with a step size of 1 and corresponding 384 bias terms. Then, the obtained result is calculated using the ReLU activation function.

The fourth layer is a convolutional layer. The 384 which size is $3 \times 3 \times 384$ convolution kernels with a step size of 1 and the corresponding 384 bias terms are used to convolute the results of the third layer. Then, the obtained result is calculated using the ReLU activation function.

The fifth layer is a convolutional layer. The 256 which size is $3 \times 3 \times 384$ convolution kernel with a step size of 1 and a corresponding 256 bias term to convolute the result of the fourth layer. Then, the obtained result is calculated using the ReLU activation function.

The sixth layer is the pooling layer, and the result of the fifth layer is the most pooled.

The seventh layer is the fully connected layer, and the pooling result of the sixth layer is expanded into 4096 neurons for linear calculation:

$$f = Wx + b$$

W is the weight of the model obtained through training, b is the bias term, and x is the output result of the previous layer. After the result is obtained, a dropout [21, 22] operation with a probability of 0.5 is performed. Dropout is an effective regularization technique. The basic idea is to improve the generalization of neural networks by preventing feature detectors from working together.

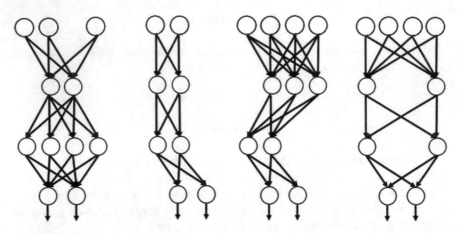

Fig. 6 The process of dropout randomly discarding nodes to form different networks

There is also an interesting phenomenon in dropout. Because he has the probability of discarding the characteristics of neuron nodes, when we define a network structure and discard the neurons through dropout, the model may have some subtle changes in each structure during training. Ultimately, the model will choose a model structure that is more suitable for this problem within the scope of the defined structure (Fig. 6).

The above figure lists some possibilities. These four structures may appear in the dropout operation, and the weight parameters are shared between the four structures. So don't worry about the extra workload when you train. At the same time, dropout technology with Maxout [23] activation function will have better results. Maxout activation function:

$$h(x) = \max_{j \in [1,k]} z_{i,j}$$

$z_{i,j} = x^T W \ldots_{ij} + b_{ij}, W \in R^{d \times m \times k}$. This is because the Maxout activation function is a locally linear function. Dropout works better with linear activation functions.

The eighth layer is the fully connected layer, and the pooled result of the seventh layer is expanded into 4096 neurons, and linear calculation is performed, and then a dropout operation with a probability of 0.5 is performed. Get the calculation result.

The ninth layer is the Softmax layer, using the Softmax function:

$$\text{soft} \max(x_1, x_2, \ldots, x_n) = \frac{e^{x_i}}{\sum_{i=1}^{n} e^{x_i}}$$

n is the number of types and the Softmax value is the probability of each type; obviously the sum of all Softmax values is 1. A Softmax function is used to generate a label distribution containing 4 categories. Cross entropy:

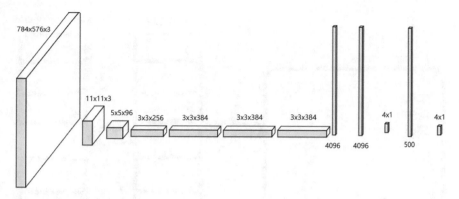

Fig. 7 The structure of the entire model

$$loss = \sum_{i=1}^{n} p(x_i) \log(q(x_i))$$

$p(x_i)$ is the true probability of x_i and $q(x_i)$ is the probability that x_i is calculated by the model. Cross entropy measures the distance between two distributions. The closer the actual distribution to the predicted distribution, the smaller the cross entropy. Finally, we use Adam optimized gradient descent to solve this model. When the parameters converge, we can get the preliminary model.

The tenth layer is a simple fully connected layer with 500 neurons, making a simple nonlinear transformation of the results from the previous models.

The eleventh layer is also a simple fully connected layer with only four neurons, making a simple nonlinear transformation of the results from the tenth layer.

The twelfth layer is a Softmax layer that is similar to the previous Softmax layer function, giving the final probability of each type.

The entire model is shown below (Fig. 7).

3.3 Train

We build a convolutional network model using the tensorflow-gpu-1.12.0 framework. The GPU is Titan. The CPU is Xeon6130. In the training process, we divided the entire data into three equal parts, and trained three different models. The data was divided into training set and verification set according to the ratio of 7:3, and each 16 pictures were used as a training batch. There are 12,000 batches. According to this, each batch is input to the convolutional neural network for training. Considering that I/O operations take longer than decoding and matrix operations, once all training data is not read into the computer memory and in order not to waste GPU resources, we use multi-threaded operation to train the model, as shown below (Fig. 8).

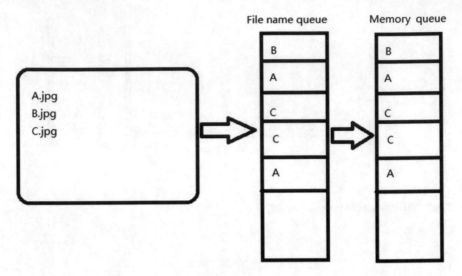

Fig. 8 The process of the multi-threaded operation

We use a thread to read the file name into the file name queue, and the decoder thread reads the image into the memory queue through the file name queue. This will speed up the training. So that the CPU does not need to wait for I/O operations, the GPU can always do matrix operations without waiting for the data to be decoded in the CPU.

4 Experiments and Results

4.1 Accuracy

We use accuracy to judge the results of the model. First, from the network structure of the following figure, we calculate the loss function and then calculate the quasi-curvature when the current loss function value (Fig. 9).

$$Precision = \frac{correctly\ classified\ post}{classified\ post} \times 100$$

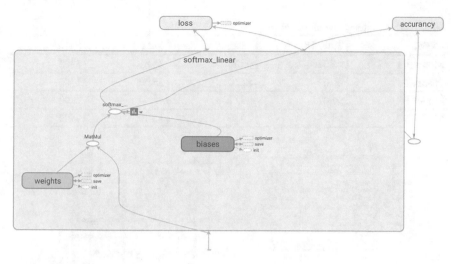

Fig. 9 The process of get the accurancy

4.2 Experimental Process and Details

First, we read the data in bulk through a multi-threaded reader. Multiple images as shown below are used as input to the model (Fig. 10).

Then, we noticed that the soft sand in the sample and the compacted soil had rock interference. We initially tried to classify the samples into six categories, soft rock with more rocks, soft sand with less rocks, compacted soil with more rocks, compacted soil with less rocks, rocky terrain and concave land.

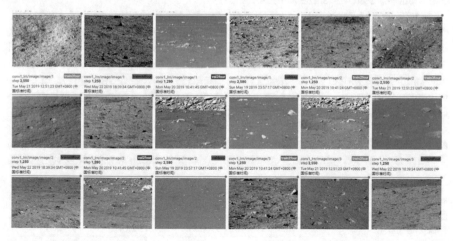

Fig. 10 The samples of one training batch

Table 2 The accuracy rate of the submodel

Model	Train set	Test set	Val set	STEP
Model-1-1	96.01	90.17	43.45	1300
Model-2-1	89.02	81.57	35.07	1300
Model-1-2	97.07	89.88	43.94	2600
Model-2-2	88.33	79.93	36.19	2600

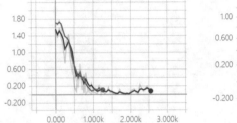 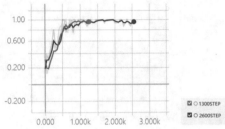

Fig. 11 The loss (left) and the accurancy (right) on training data with 2600 epochs

We divided the marked data into three parts. We take two of them and put each picture into the network and train it. The accuracy rate is as follows (Table 2).

In terms of model training, Model-1-1 and Model-1-2 are obtained by performing 1300 epochs and 2600 epochs on the same data. We observe the loss function graph and the accuracy graph during the training process (Fig. 11).

We found that the model has converged when the model has been trained 1300 epochs. So we don't have to do extra training. The processes of training Model-2-1 and Model-2-2 are the same as those.

In terms of accuracy, we found that the model has a poor accuracy for new data classification, which means that the model has a very low generalization ability. Further, we analyze the reasons and find that in the classification results, the model has fuzzy judgment ability on the number of rocks. The essential reason is that in the sample, the judger's judgment on the rock is also fuzzy. We believe that this boundary blur condition should be removed, so we decided to divide the book into four categories, soft sand, compacted soil, rocky terrain and concave land (Table 3).

We divided the marked data into three parts. Put each picture into the network and train 1300 epochs. The accuracy rate is as follows (Fig. 12).

The loss function value and accuracy of each training batch are as follows (Fig. 13).

Table 3 The accuracy rate of the submodel witch training 1300 epochs

Model	Train set	Test set	Val set
Model-1	97.606	91.07	70.44
Model-2	92.50	87.17	72.61
Model-3	89.74	89.30	76.94

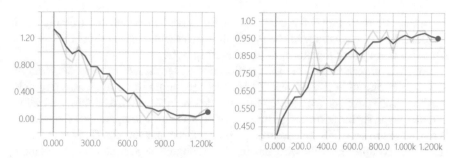

Fig. 12 The loss (left) and the accurancy (right) on training data with 1300 epochs

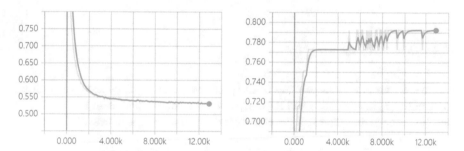

Fig. 13 The loss (left) and the accurancy (right) of the entire model

We found that the model converges with higher accuracy and lower loss function values.

Finally, we use the idea of ensemble learning to fuse three models and then train a simple neural network based on the results of the three models. We take its output as the final output. This process is similar to a voting process, but the weight of each vote is non-linear and it is a neural network that needs to be trained.

The loss function value and accuracy of each training batch are as follows.

Obviously, we see that as the training batch increases, the loss function value decreases and the quasi-curvature rises. It is basic convergence to 12,000 times. Finally, we see that the average quasi-curvature is stable at 80%. The average accuracy of the previous method was 78.4% [10, 11].

Compared to a single model, the overall model, the overall model will be more accurate than a single model (Table 4).

Table 4 The accuracy rate of the entire mode

Model	Train set	Test set	Val set
Model-1	97.61	91.07	70.44
Model-2	92.50	87.17	72.61
Model-3	89.74	89.30	76.94
Model-x	81.25	80.17	80.01

Among them, Train is the accuracy of the model in which the training set is divided, Val is the accuracy of the model in which the validation set is divided, and All is the accuracy of the model in the overall sample.

For the compacted soil, the recognition rate of our model reached 87.96%, the recognition rate of soft sand was 79.1%, the recognition rate of rock topography reached 71.2%, and the recognition rate of concave land reached 33.4%. We believe that the reason for the poor recognition rate of concave land compared to other terrains is that there are only 368 concave topographic maps in our sample. Conversely, there are 2,765 compacted soils with the highest recognition accuracy in our sample, indicating that the model requires a large number of samples to learn, in order to truly learn to extract features and correctly classify the terrain.

4.3 ROC Curve

There are four situations existing in the prediction results. They are true positive, false positive, true negative and false negative. We abbreviated them to TP, FP, TN, and FN respectively represent the corresponding number of samples, obviously:

$$TP + FP + TN + FN = \text{Total number of samples}$$

as the picture shows (Fig. 14).

We define two formulas:

$$TRP = \frac{TP}{TP + FN}$$

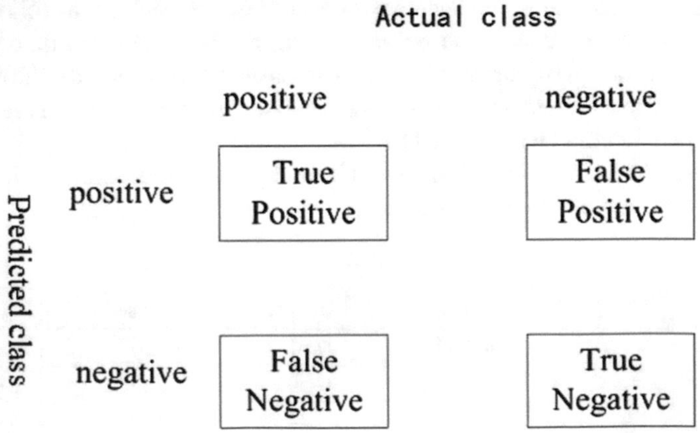

Fig. 14 The confusion matrix

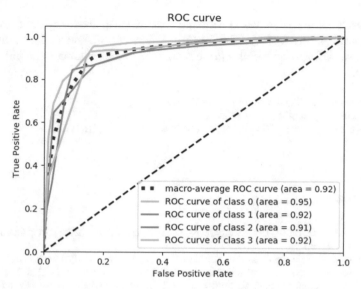

Fig. 15 The ROC curve

$$FRP = \frac{FP}{TN + FP}$$

Make a ROC [20–26] image as follows (Fig. 15).

By calculating the data of TRP, TP, FN, TN, FRP, we can get m test samples for each category and the probability matrix for that category. Therefore, according to each column in the probability matrix label matrix, an ROC curve is drawn. In this way, a total of n ROC curves can be drawn. Finally, the average ROC curve is obtained by averaging the n ROC curves.

5 Conclusions and Future Work

The original data of this study was from the Chang'e III's camera. We apply convolutional neural networks and integrated learning to the identification of lunar surface terrain, and propose a multi-network integrated learning model to predict the terrain categories within the field of view and achieve a better accuracy. Although, neural networks do not require as much computational complexity and hardware support as backpropagation in feedforward computational reasoning. However, in the current general view, the deeper the hierarchy, the more parameters the network tends to have better results. However, this is not sufficient for the calculation of speed, hardware requirements or power supply requirements for machines in many special working environments (such as lunar calendars). Fortunately, people are no longer relying on experience in network design. The development of AutoML [27] has led people to use machines to design neural networks that are small and comparable to human experts, such as NasNet [9], PNasNet [28], MnasNet [29], and so on. In the future research, we will also move closer to the design of this fully automated neural network structure.

Acknowledgements Thank you for your guidance.
Thanks for the support of the research team.
Thanks to the support of the National Natural Science Foundation (41671402).

References

1. Serikawa S, Lu H (2014) Underwater image dehazing using joint trilateral filter. Comput Electr Eng 40(1):41–50
2. Lu H, Li Y, Mu S, Wang D, Kim H, Serikawa S (2018) Motor anomaly detection for unmanned aerial vehicles using reinforcement learning. IEEE Internet Things J 5(4):2315–2322
3. Lu H, Li Y, Chen M, Kim H, Serikawa S (2018) Brain intelligence: go beyond artificial intelligence. Mob Netw Appl 23:368–375
4. Lu H, Wang D, Li Y, Li J, Li X, Kim H, Serikawa S, Humar I (2019) CONet: a cognitive ocean network. IEEE Wirel Commun. In Press
5. Lu H, Li Y, Uemura T, Kim H, Serikawa S (2018) Low illumination underwater light field images reconstruction using deep convolutional neural networks. Fut Gen Comput Syst 82:142–148
6. Deng J, Dong W, Socher R et al (2009) ImageNet: a large-scale hierarchical image database. In: 2009 IEEE computer society conference on computer vision and pattern recognition (CVPR 2009), 20–25 June 2009, Miami, Florida, USA. IEEE
7. Krizhevsky A, Sutskever I, Hinton G (2012) ImageNet classification with deep convolutional neural networks. NIPS. Curran Associates Inc.
8. Simonyan K, Zisserman A (2014) Very deep convolutional networks for large-scale image recognition. Comput Sci
9. Zoph B, Vasudevan V, Shlens J et al (2017) Learning transferable architectures for scalable image recognition
10. Brooks CA, Iagnemma K (2012) Self-supervised terrain classification for planetary surface exploration rovers. J Field Robot 29(3):445–468
11. Ojeda L, Borenstein J, Witus G et al (2006) Terrain characterization and classification with a mobile robot. J Robot 23(2):103–122
12. Hubel D, Wiesel T (1959) Receptive fields of single neurones in the cat's striate cortex. J Physiol 143(3):574–591
13. Hubel D (1962) Wiesel T. Receptive fields, binocular interaction and functional architecture in the cat's visual cortex 160(1):106–154
14. LeCun Y, Boser B, Denker J, Henderson D. Howard. R, Hubbard W, Jackel D (1989) Backpropagation applied to handwritten zip code recognition. J Neural Comput 1(4):541–551
15. Behnke S (2003) Hierarchical neural networks for image interpretation. Springer, Berlin, Heidelberg
16. Simard P, Steinkraus D, Platt J (2003) Best practices for convolutional neural networks applied to visual document analysis. ICDAR 2:958
17. Ahn SM (2016) Deep learning architectures and applications. J Intel Info Syst 22(2):127–142
18. Ogiela L, Ogiela MR (2012) Advances in cognitive information systems. Cognit Syst Monogr 17:1–18
19. Seide F, Gang L, Dong Y (2011) Conversational speech transcript using context-dependent deep neural networks. Interspeech

20. Zeiler MD, Fergus R (2014) Visualizing and understanding convolutional networks. In: Computer vision–ECCV 2014, pp 818–833
21. Hinton GE, Srivastava N, Krizhevsky A et al (2012) Improving neural networks by preventing co-adaptation of feature detectors. Comput Sci 3(4):212–223
22. Wan L, Zeiler M, Zhang SX et al (2013) Regularization of neural networks using dropconnect. In: Proceedings ICML, pp 2095–2103
23. Goodfellow IJ, Wardefarley D, Mirza M et al (2013) Maxout networks. In: International conference on machine learning
24. Kingma DP, Ba J (2015) Adam: a method for stochastic optimization. In: International conference on learning representations
25. Fawcett T (2006) An introduction to ROC analysis. Pattern Recogn Lett 27(8):861–874
26. Hand DJ, Till R (2001) A simple generalisation of the area under the ROC curve for multiple class classification problems. Mach Learn 45(2):171–186
27. Kaul A, Maheshwary S, Pudi V et al (2017) AutoLearn—automated feature generation and selection. In: International conference on data mining, pp 217–226
28. Liu C, Zoph B, Neumann M et al (2018) Progressive neural architecture search. In: European conference on computer vision, pp 19–35
29. Tan M, Chen B, Pang R et al (2018) MnasNet: platform-aware neural architecture search for mobile. Comput Vis Pattern Recogn 2820–2828

An Automatic Evaluation Method for Modal Logic Combination Formula

Xueqing Li, Zhujun Wang, Yu Xia, and Yi Jiang

Abstract Modal logic is widely used in the knowledge representation of intelligent knowledge systems, but the strict deduction of dependency axioms and inference rules of compound formulas in the world hinders its application in practical systems. This paper presents an automatic evaluation algorithm of modal logic compound formula based on Haskell functional programming to solve the problem of evaluating modal logic compound formula in the possible world. In this paper, an M^+ model is presented, and on this basis, the automatic evaluation of the compound formula in the possible world is realized. Experimental and case results verify the effectiveness and accuracy of the proposed algorithm, and its automatic solution obviously reduces the manual dependence.

Keywords Modal logic · Possible worlds · Evaluation algorithm · Knowledge representation · Haskell functional programming

1 Introduction

Modern modal logic appeared in the 1910s. Later, due to the diversified development of modern modal logic research, in addition to philosophical analysis, the application of modal logic has increasingly become a study of modal logic. Important directions include mathematics [1], social sciences [2], economics [3], engineering [4], and computer science [5, 6]. Modern modal logic is applied in the formal representation of subject knowledge of intelligent systems, knowledge communication and knowledge reasoning, description or verification protocols, and design of knowledge-based support decision systems [7]. The XYZ/E (Sequential Logic Language) based on modal logic has been successfully applied to the hardware system [8], the behavior description and verification of the hybrid system [9] and the interpretation of program semantics. Model detection techniques proposed by American scientists CLarke and

X. Li · Z. Wang · Y. Xia · Y. Jiang (✉)
School of Information Engineering, Yangzhou University, Yangzhou, China
e-mail: jiangyi@yzu.edu.cn

© Springer Nature Switzerland AG 2021
H. Lu (ed.), *Artificial Intelligence and Robotics*,
Studies in Computational Intelligence 917,
https://doi.org/10.1007/978-3-030-56178-9_17

Emerson [10] use formulas in modal logic to describe the properties of the system. In the multi-agent system, the description of knowledge and beliefs by modal logic can solve the ambiguous problem of conscious concepts such as "belief" and "wish" in first-order logic, thus having clearer expressive power and more effective. The form of reasoning [11] can be applied to a given set of tasks and agents. When assigning corresponding tasks to different agents, it needs to be applied to the corresponding policies. The choice and representation of the strategy can be formalized into the true and false value of the compound formula in the modal logic. Modal logic has moved from philosophical fields to more discipline applications.

Compound formula evaluation is a very important branch of modern modal logic. The artificial evaluation method based on axioms and deduction rules, if faced with a relatively simple state space at this time, the computational difficulty of manually calculating the true and false values of the compound formula can be controlled, when faced with the possible world model of the complex space state, because of the need Considering that all the possible worlds involved in the current propositional formula and the truth values of the propositions in different reachable relationships are different, the general artificial compound formula is difficult to evaluate and the correct rate is low.

In order to improve the efficiency and correctness of the modal logic compound formula evaluation. This paper proposes to implement the automatic evaluation of compound formulas in modern modal logic in every possible world based on Haskell functional programming method. Haskell functional programming is different from conventional imperative programming. Functional programming abstracts the computational process into expression evaluation. The expression consists of pure mathematical functions, which is closely related to the formulas in modal logic, which is more conducive to modal logic. Formal expression of axioms and deduction rules. First, we build a model of modal propositional logic. Then we will give each possible world in the model, and fix the assignment function of all primitive propositions according to the assignment function in the model. Different truth value definitions determine the true value of the compound formula in each possible world. Then we propose a compound formula automatic evaluation algorithm based on Haskell functional programming. Finally, we demonstrate the application of this method through an experiment.

The organizational structure of this paper is as follows: In the second part, we discuss the related concepts of modal logic compound formula and common evaluation methods; the third part gives definitions of concepts such as propositions and possible worlds in modal logic, and introduces the model. The model of state proposition logic gives the relevant theorems and deduction rules. The fourth part introduces the automatic evaluation algorithm of modal logic compound formula. The fifth part discusses the application of this method under a specific case model. The final part is the conclusion and future work.

2 Related Work

The description of knowledge and formal reasoning are important issues in the study of intelligent knowledge systems. In the traditional modal logic, the method of calculating the true and false values of the compound formula has artificial reasoning and resolution method. Manual Deduction Method Step 1: Define a non-empty language set K that contains the context of all possible situations. If there is a temporal structure such as "always" or "every" in the situation, then the context set contains the description of the time. If there is a modal structure such as "necessary" and "possible", then we will confirm the context as considering all possible situations. The second step is to formalize the relevant strategy or knowledge into a primitive proposition, and superimpose the modal operator and the logical conjunction on the primitive proposition to represent the corresponding modal proposition. Step 3: Estimate the composite proposition truth value for each possible world based on the specific context k (taken from the context set K). The disadvantage of this method is that it is difficult to control the calculation difficulty of manually calculating the true and false values of the compound formula in the face of a relatively complicated state space.

The semantic tableau method proposed by Beth and Hintikka extends the formula construction set, which is generally applicable to the formula reasoning of logic systems. On the basis of the classical semantic tableau, Liu Quan and Sun Jigui applied non-classical logic automatic reasoning methods that reinterpreted symbols, extended symbols, modified closed rules, changed tree extension methods, added edge information, and used dual trees. In tableau, the efficiency of machine inference execution in non-classical logic formulas is improved [12].

Zhang Jian proposed the translation method of modal logic reasoning [13]. He proposed to translate the modal logic formula into a classical logic formula according to certain rules, and then use the traditional theorem prover to reason. This method theoretically maintains the decidability of the formal propositional modal logic.

In the research of modal logic formula derivation, Liu Lei, Wang Qiang and Lu Shuai proposed to extend the Tableau method in classical modal logic to FPML, and proposed a reduction strategy based on FPML Tableau rules and fuzzy assertion set [14]. On this basis, the definitions of inconsistency and inconsistent estimates in FPML are given. Finally, the CID of the FPML consistency detection method TFPML and the fuzzy assertion set based on the Tableau method is given, which proves the validity and correctness of the method in the modal logic formula derivation.

When Zhou Juan and Li Chao studied the modal inference problem, the method of deductive reasoning in modal logic was transformed into modalsimic multi-valued logic with the necessary formal methods, and then Lucasivi The multi-valued logic is converted to Boolean logic [15]. The results show that compared with other methods, the inference engine has universality, computational simplicity, and unreasonable application of inference rules in modal logic.

In addition to this, another common method is the resolution method. Sun Jigui and Liu Xuhua proposed label-based modal inference [16], which overcomes the excessive symbolic redundancy of L. Farinasdel Cerro's propositional modality method, and adds a formula for the reduction of a formula under two possible sub-constraints. The modal Associative Method proves the reliability and completeness of the labeling modality. This new modal categorization method is almost 10 times faster than the modality of P. Enjalbert et al.

Pan Weimin and Chen Tuyun proposed a new modal regression [17]. They gave a definition of a standard clause, on the basis of which it defines the form of the clause set of the propositional modal logic system S5. It is proved that the necessary and sufficient condition for the unsatisfiable arbitrary mode S5 clause set is that it can be attributed to the empty clause in its form of conclusion, which reduces the formula redundancy.

The application of the resolution method to classical logic is very effective [18]. However, there are difficulties in applying modal logic. Because modal logic belongs to a kind of non-classical logic, the resolution method is closely related to the conjunction paradigm, but the non-classical logic formula is difficult to transform into the form of the forefront paradigm required by the resolution method. More importantly, the different systems that are required for different logic systems have different resolution systems [19]. Therefore, we propose a method based on Haskell functional programming to realize the automatic evaluation of compound formulas in modern modal logic in every possible world. This method has strong versatility and intuitiveness.

3   Basic Concepts and Problem Formalization

Modal logic is a complex subject. This section introduces the definition of modal logic, modal operators and modal propositions, and models them according to relevant knowledge points to prepare for the fourth algorithm.

3.1   Modal Logic Related Concepts

The concept of the possible world comes from GW Leibinz's idea of non-contradictory possibility [20]: "As long as the combination of the situation of the thing and the situation of the transaction does not push the logical contradiction, the combination of the situation of the thing or the situation of the thing is possible". In this thought, Leibniz proposed: "The world is a combination of possible things. The real world is a combination of all possible things that exist (the most abundant combination). There may be different combinations of things, which will form many The possible world." Therefore, the possible world is a combination of possible things that are used to express modal assertions.

There are three types of modal words in modal logic: the first type of words that represent the "state" of things such as "inevitable" and "possible"; the second type of words that express the "modality" of a character such as "know" and "Believe in" etc.; the third category of words that represent the "temporal" of the process such as "future", "consistent" and so on. The combination of modal words and meta-propositions forms a new modal proposition.

Propositional form: $M_a p$, where a is the cognitive subject agent, M is a cognitive modal word, and p is an arbitrary proposition.

The context involved in the concept refers to the state and state of the person or thing. Perhaps the context of the world is closely related to the type of modal words. If the modal word is a word that expresses tense, such as "future" and "consistent", the context is a description of the moment of time.

A model M is defined as follows:

1. A non-empty context set K;
2. A binary relationship R on K is called a reachable relationship;
3. Value function V, for each context $k \in K$, each primitive proposition p is assigned a true value $V_k(p)$;

The collection of all contexts is called the mirror set (represented by K). The calculation of the true and false values of the modal proposition depends not only on the current given context, but also on other contexts, which is related to the specific meaning of the modal words. This means that an evaluative function that assigns a true value according to a specific context k (taken from the vocal set K) is used instead of a simple formula to assign an artificial evaluation of the absolute true value.

The reachable relationship is a binary relationship (represented by R) on the set of K, and the context k' associated with the estimate in a context k is said to be reachable from k.

3.2 Modal Operators and Modal Propositions

$\Diamond$: *possible operator*; $\Box$: *Inevitable operator*; *p* : *an arbitrary proposition*

$\Diamond p$: *p is possible*; $\Box p$: *p is Inevitable*;

All propositions containing modal operators are collectively referred to as modal propositions. We divide it into the following four categories:

Proposition type	Symbol	Meaning	Example
Necessary affirmative proposition	$\Box p$	A proposition that determines the certain existence of things	The organism must undergo metabolism
Necessary negative proposition	$\Box \neg p$	A proposition that determines the Inevitabe existence of things	Objective law must not be transferred by human will
Probable affirmative proposition	$\Diamond p$	A proposition that determines the probable existence of things	Long-term insomnia may get sick
Probable negative proposition	$\Diamond \neg p$	A proposition that determines the probable non-existent things	People can't live forever

Hypothesis the proposition form have **p** and ¬**p**, inevitably = $\Box$, maybe = $\Diamond$ then:

Axiom 1:$\Box \neg p \leftrightarrow \Diamond p$
Axiom 2:$\Box p \leftrightarrow \Diamond \neg p$

These two axioms behave as the relationship between the "necessary" operator and the "possible" operator, which can be visually seen through the modality. Through these axioms, the two operators can be freely converted. This serves as the basis for deriving the evaluation of the modal logic compound formula.

3.3 Model Assignment Function

Let M be a model of the possible world set W, R and V are the reachable relationship and the assignment function respectively. Given the true value of p in w W, the correlation theorem is as follows:

1. $V_{M,w}(p) = V_W(p)$,p is the primitive proposition;
2. $V_{M,w}(\neg p) = true$, if and only if $V_{M,w}(p) = false$;
3. $V_{M,w}(\Box p) = true$, if and only if $\forall w' \in W$ $satisfy$ wRw', $then$ $V_{M,w}(\Box P) = True$;
4. $V_{M,w}(\Diamond p) = true$, if and only if $\exists w' \in W$ $satisfy$ wRw', $then$ $V_{M,w}(\Box P) = True$.

We can clearly see the "necessary" modal operator in table above, which means that the primitive proposition p is true in all reachable worlds, defining the "possible" modal operator in 4, meaning that the primitive proposition p At least one of the worlds is true.

3.4 The Model

In the research process of this paper, we fill the model with specific data, and the filled model M^+ is as follows:

1. A possible world of non-empty context sets W = w1, w2, ..., wn has n possible worlds;
2. A binary set R on W = (w1, w2), (w2, w2), ..., (wi, wj), ..., (wk, wn), called the reachable relationship set. Where (wi, wj) means that wi can reach wj, and if i = j, it means that the possible world is self-achievable.
3. The value function V, for each world $w \in W$, each primitive proposition p is assigned a true value $V_W(p)$. When the model is established, we will fix the assignment function V, that is, the true value of the proposition in all possible worlds is $V_{Wi}(p) = true$ $V_{Wj}(p) = false$ indicates that the primitive proposition has a value of true under the possible world wi and a value of false under wj.

We will discuss the true value of $\Box p$ and $\Diamond p$ and the overlapping modal operators $\Box \Diamond p$ and $\Diamond \Box \neg p$ in each possible world. We need to set up an efficient algorithm to quickly and accurately obtain the true and false values of the modal logic compound proposition in the case of an increase in the number of modal operator superpositions and an increase in the complexity of the world state space.

4 Automatic Formula Algorithm for Complex Formulas

Algorithm 1 solves the true value of the proposition under the action of the possible operator in the possible world of the target. First, we use a for loop to traverse the reachable relationship model to find the reachable world of the target possible world. We abstract it into a function filtrate using Haskell functional programming as a tool. The function filtrate::String->Model->[String] indicates that the function takes a string type and a Model type as input, and a String list as output. String represents the possible world of the target, and Model is the set of reachable relationships. Next, using a double loop, the inner loop traverses to find the true value of the proposition in the target possible world proposition, and the design function finds the corresponding true value of the possible world. The function findbool::String->Proposition->Bool indicates that the input is a character. The string and a collection of type Proposition, the output is the Bool value, and the function is used to find the true value of the proposition in the possible world of String. The outer loop is to take all the true values of all possible worlds by traversing the possible world, and judge the truth value according to the definition. The design function posvfun::Proposition->Model->String->Bool means that the Proposition type binary group, the Model type binary group, and the String type representing the target possible world are taken as inputs, the truth value is judged, and the Bool type value is output.

Algorithm 2 and Algorithm 1 are similar, and the filtrate function is the same as the findbool function. When traversing the possible world, the design function necvfun::Proposition->Model->String->Bool is designed according to the difference between the possible operator and the inevitable operator in the value of the true value. Although the input and output types are the same, But the design procedure is different. In the outer loop of the double loop, the number of possible true world values is true, and the non-algorithm is a true value that determines whether there is true.

Theorem 1 $Let W = w1, w2, ...wn R = (w1, w2)...(wi, wi)...(wk, wn);$ $V_{M,wi}$ $(p) = true;$

$Then: V_{M,w} (\Diamond p) = true$

Theorem 1 states: When the possible world is self-sufficient and the primitive proposition is true in the possible world, the proposition under the action of the operator may be true in the possible world.

Algorithm 1: POSOPERATOR(WN,PRO,MODEL)

Input: WN,PRO,MODLE
Output: Bool
1 $List \leftarrow \Phi$
2 **for** $\forall(pre, suc) \in MODEL$ **do**
3   **if** $pre{==}WN$ **then**
4 | $List \leftarrow List \cup \{suc\}$
5   **end**
6 **end**
7 $flag \leftarrow false$
8 **for** $\forall(Pw, Bo) \in PRO$ **do**
9   **for** $\forall i \in List$ **do**
10   **if** $i{==}Pw$ and $Bo{==}true$ **then**
11 | $flag \leftarrow true$
12   **end**
13   **if** $flag{==}true$ **then**
14 | return true
15   **end**
16   **end**
17 **end**
18 return $false$

Algorithm 2: NECOPERATOR(WN,PRO,MODEL)

Input: WN,PRO,MODLE
Output: Bool
1 $List \leftarrow \Phi$
2 **for** $\forall(pre, suc) \in MODEL$ **do**
3   **if** $pre{==}WN$ **then**
4 | $List \leftarrow List \cup \{suc\}$
5   **end**
6 **end**
7 $flag \leftarrow false$
8 **for** $\forall(Pw, Bo) \in PRO$ **do**
9   **for** $\forall i \in List$ **do**
10   **if** $i{==}Pw$ and $Bo{==}true$ **then**
11 | $flag \leftarrow flag + 1$
12   **end**
13   **end**
14 **end**
15 **if** $flag{==}length\ of\ List$ **then**
16 | return true
17 **end**
18 return $false$

5 Experimental Results and Analysis

In order to prove the correctness of the algorithm and theorem proposed in the previous section, we first construct a modal logic model of a multi-possible world. On this basis, the validity and accuracy of the proposed algorithm are verified.

5.1 Experimental Model

According to the experimental model in 3.3, the M+ model is given, in which the world of w2 is self-achievable, and the correctness of Theorem 1 is proved by the automatic evaluation of the formula under the possible operator of w2 world. The model is as follows (Fig. 1).

5.2 Experimental Results and Verification

Verification: In w1,
Input: model=[("W1", "W2"),("W2", "W2"),("W2", "W3")];
   proposition=[("W1",True),("W2", True),("W2", False)];
Output: posvfun proposition model "W1"=true;
   nesvfun proposition model "W1"=true;
 Because $w1Rw2$, $V_{M,w2}(p) = true$,
  So $V_{M,w1}(\Diamond p) = true$,
 And because $w2$ is the only $w1$ reachable world;
  So $V_{M,w1}(\Box p) = true$.
Verification result: The experimental result is correct.

Verification: In w2,
Input: model=[("W1", "W2"),("W2", "W2"),("W2", "W3")];
   proposition=[("W1",True),("W2", True),("W2", False)];

Fig. 1 M^+ Model diagram

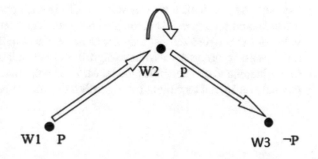

Output: posvfun proposition model "W2"=true;
      nesvfun proposition model "W2"=false;
  Because p is ture in $w2$ itself and $w2Rw2$,
    So $V_{M,w2} (\Diamond p) = true$,
And because $w2$ is the only $w1$ reachable world;
    So $V_{M,w2} (\Box p) = false$.
Verification result: The experimental result is correct.

Verification: In w3,
Input: model=[("W1", "W2"),("W2", "W2"),("W2", "W3")];
          proposition=[("W1",True),("W2", True),("W2", False)];
Output: posvfun proposition model "W3"=false;
      nesvfun proposition model "W3"=true;
  Because $w3$ simply does not reach the world,
  So p is true in all reachable worlds (even if all reachable the world is not there at all);
    So $V_{M,w3} (\Diamond p) = false, V_{M,w3} (\Box p) = true$;
Verification result: The experimental result is correct.

5.3 Experiment Analysis

By comparing the experimental results with the verification results, we can clearly see that the modal logic compound formula evaluation algorithm based on Haskell functional programming is characterized by high accuracy and high efficiency. In the M^+ model, it is possible that the world w2 is self-achievable, and the primitive proposition p is true at w2, and it can be deduced that the proposition under the action of the possible operator is always true in the possible world, and the experimental results prove The correctness of this theorem. In addition, we use this algorithm to perform the true value calculation under the action of overlapping modal operators, and the experimental results obtained are also correct and efficient. Because in the modal propositional logic, the propositional form has $\Box p$ and $\Diamond p$, and also $\neg p, \Box \Diamond p$, $\Diamond \Box p$ etc. These can also be calculated using the above algorithm. Taking $\Diamond \Box p$ as an example, to calculate the true value of the proposition $\Diamond \Box p$ in the possible world Wi, we need to calculate the true value of $\Box p$ in each possible world, then treat it as a whole, and then give it according to posvfun. The method of calculating the true value of the proposition that may act as an operator in the possible world Wi. It is only necessary to master the above algorithm to calculate the true value calculation on the larger and more complex model M. The compound overlap of modal operators can only be solved by calling the algorithm multiple times.

6 Conclusion and Prospect

The evaluation of compound formulas in modal logic in different possible worlds is a big problem. The key task is the design and implementation of model establishment and evaluation algorithms. At present, there are still difficulties in how to construct efficient and concise models and formal expression of complex propositional formulas, which needs further study. The modal logic compound formula evaluation algorithm based on Haskell functional programming proposed in this paper, after relevant experimental research, can be seen from the results, the accuracy of the modal logic compound formula evaluation result is obtained with the participation of Haskell functional programming. Guarantee, improve the calculation speed, and in the evaluation of the problem of the more complex overlapping modal operator, the accuracy and portability are effectively guaranteed.

Acknowledgments This work was supported by the National Nature Science Foundation of China under Grant 61872313, the key research projects in education informatization in Jiangsu province (20180012), in part by the Postgraduate Research and Practice Innovation Program of Jiangsu Province under Grant KYCX18 2366, and in part by the Yangzhou Science and Technology under Grant YZ2017288, YZ2018076, YZ2018209, and Yangzhou University Jiangdu Highend Equipment Engineering Technology Research Institute Open Project under Grant YDJD201707, and Jiangsu Practice Innovation Training Program for College Students under Grant 201811117029Z.

References

1. Liangkang N (2017) Mathematical philosophy of phenomenology and modal logic of phenomenology-from the perspective of Husserl and Becker's thoughts. Acad J 1:11–27
2. A Beneficial probe into the applied research of logic—comment on inevitability, possibility and contradiction—an analysis of Jon Elster's Logic and Society. J Henan Financ Tax Coll 2017(03):2
3. Tiede C, Tiampo K, FernNdez J et al (2010) Fuzzy logic model for the determination of physical reliability of volcanic sources. J Am Ceram Soc 86(4):533–533
4. Liau CJ, Lin IP (1992) Quantitative modal logic and possibilistic reasoning. In: European conference on artificial intelligence
5. Fitting M, Mendelsohn RL (2006) First-order modal logic. Bull Symb Logic 8(3):549–620
6. Pauly M (2002) A modal logic for coalitional power in games. J Logic Comput 12(1):149–166(18)
7. Zhang H (2003) Research on cotton management decision support system based on knowledge model. Nanjing Agricultural University
8. Shilov NV, Yi K (2001) How to find a coin: propositional program logics made easy
9. An Z, Zhisong T (2000) Hybrid system based on XYZ/E. J Softw 11(1):1–7
10. Clarke EM (1981) Design and synthesis of synchronization skeletons using branching time temporal logic. LNCS 131
11. Chen ZY, Huang S, Han L (2013) Application of modern modal logic in computer science. Comput Sci 40(S1):70–76
12. Quan L, Jigui S (2002) Semantic tableau method for nonclassical logic. Comput Sci 29(5):72–75
13. Jian Z (1998) Translation method of modal logical reasoning. J Comput Res Dev 5:389–392

14. Liu L, Wang Q, Lü S (2017) Tableau method of fuzzy propositional modal logic. J Harbin Eng Univ 2017(6)
15. Zhou J, Li C (2015) Modal logic inference engine based on Lucasiwitz multi-valued calculus. J Hubei Univ Natl Nat Sci Edn 3:285–289
16. Jigui S, Xuhua L (1996) Marking modal resolution reasoning. J Softw A00:156–162
17. Weimin P, Tuyun C (1997) A new modal regression. Chin J Comput 20(8):000711–717
18. Robinson JA (1965) A machine-oriented logic based on the resolution principle. J ACM 12
19. Cerro LFD (1982) A simple deduction method for modal logic. Inf Process Lett 14(2):49–51
20. Changle Z (2001) Introduction to cognitive logic. Tsinghua University, Beijing, pp 7–9

Research on CS-Based Channel Estimation Algorithm for UWB Communications

Wentao Fu, Xingbo Dong, Xiyan Sun, Yuanfa Ji, Suqing Yan, and Jianguo Song

Abstract It is difficult to estimate the UWB channel for traditional algorithms, and the compressed sensing greedy algorithm must solve the problem of UWB channel reconstruction based on the sparsity. Based on the subspace tracking sparse reconstruction algorithm, an adaptive sparsity subspace tracking (ASSP) algorithm is proposed. The algorithm obtains sparsity according to the principle of minimum residua, accurately estimates the ultra-wideband channel at a lower sampling rate, and gives the empirical principle of setting the initial value of sparsity. The simulation results show that the ASSP algorithm has better reconstruction probability than the traditional sparse reconstruction algorithm under low SNR and the same measurement value, and can obtain better mean square error in channel estimation performance.

Keywords UWB · Compressed sensing · Sparse · Channel estimation

1 Introduction

UWB communication is one of the key technologies for short-range communication. UWB has the advantages of strong penetrability, high positioning accuracy, low power, high transmission rate and strong anti-interference ability. It is widely

W. Fu · X. Dong · X. Sun (✉) · Y. Ji (✉) · S. Yan · J. Song
Guangxi Key Laboratory of Precision Navigation Technology and Application, Guilin University of Electronic Technology, Guilin 541004, China
e-mail: sunxiyan1@163.com

Y. Ji
e-mail: jiyuanfa@163.com

X. Sun
Guangxi Experiment Center of Information Science, Guilin 541004, China

Y. Ji
National & Local Joint Engineering Research Center of Satellite Navigation and Location Service, Guilin 541004, China

© Springer Nature Switzerland AG 2021
H. Lu (ed.), *Artificial Intelligence and Robotics*,
Studies in Computational Intelligence 917,
https://doi.org/10.1007/978-3-030-56178-9_18

used in various fields such as through-wall radar, indoor positioning, and high-speed wireless LAN [1−6]. UWB uses nanosecond or sub-nanosecond narrow pulses with bandwidths up to several GHz. According to the Shannon-Nyquist sampling theorem, the sampling rate is at least twice the bandwidth in order to sample and receive the signal without distortion. Therefore, taking a 0.7 ns UWB pulse as an example, the sampling rate needs to be as high as several tens of GHz. The higher sampling rate puts higher requirements on the hardware technology, which increases the workload on the one hand and increases the cost on the other hand.

The merits of the channel estimation technique [7] determine the performance of the UWB system. However, the channel estimation has high requirements on the sampling value. How to accurately estimate the channel at a low sampling rate is an urgent problem to be solved.

Compressed sensing is a technique for extracting and reconstructing sparse signals at a lower sampling rate [8, 9], which breaks the shackles of the traditional Shannon sampling theorem. Therefore, the idea of applying compressed sensing technology to UWB channel estimation has attracted wide attention from scholars all over the world. Literature [10] focuses on the design of observation matrices and over-completed dictionary libraries. In order to avoid amplifying noise, a filter module is creatively added at the transmitting end. The reconstruction algorithm uses the DS algorithm, the base tracking noise reduction algorithm and the Orthogonal Matching Pursuit (OMP) algorithm to reconstruct the channel. The literature [11, 12] combines Bayesian theory with the theory of compressed sensing, adaptively sets hyperparameters for the reconstructed vector, and uses the maximized edge likelihood algorithm to filter and reconstruct. In [13], the UWB channel is sampled under-Nyquist sampling, and the OMP algorithm is used to estimate the sparse ultra-wideband channel. The above algorithms are all performed under the premise that the sparsity is known. In the actual UWB transmission system, the number of multipaths of the channel is often random and unpredictable. The reconstruction algorithm in [14, 15] uses the sparse adaptive matching tracking (SAMP) algorithm to reconstruct the UWB channel under unknown sparsity. However, the iterative condition is to set a fixed threshold for the residual, and the noise is randomly generated, so there is no fixed reasonable termination algorithm threshold.

This paper proposes an ASSP algorithm based on subspace tracking sparse reconstruction algorithm for UWB channel reconstruction problem with unknown sparsity. Firstly, the UWB channel estimation model is established, and then the sparsity is adjusted according to the minimum residual, and the empirical principle of initial setting of sparsity is given. Finally, the simulation is designed to verify the reconstruction performance and channel estimation performance of the proposed ASSP algorithm.

The rest of this paper is structured as follows. Section 2 discusses related work. In Sect. 3, we introduced the algorithm proposed in this paper. This section is mainly composed of the description of the adaptive sparsity subspace algorithm, which includes the UWB channel model, the principle of adaptive sparsity and the steps of the improved algorithm. Section 4 presents the evaluation of our experimental results for the MFAPCE algorithm, and Sect. 5 presents out conclusions.

2 Related Work

2.1 Compressed Sensing Theory

With the advent of the information age, the contradiction between data volume and data processing capabilities has intensified. Compressed sensing theory is proposed to solve this contradiction. It compresses and samples a sparse signal of length N into a signal of length M and processes it. The mathematical model is:

$$y = \Phi x. \tag{1}$$

$\Phi \in R^{M \times N}$ is the observation matrix $(M = N)$, M is the observation vector. $y \in R^{M \times 1}$ is the observation matrix. For a sparse signal of length N of x, the observation matrix is used to sample the signal to obtain an observation vector.

To reconstruct the signal x from the observation vector, Φ is subject to the Restricted Isometry Property (RIP) [16]. From the magnitude relationship between M and N, we can see that the Eq. (1) is a problem of solving an underdetermined equation. Reconstructing the sparse signal x requires linear optimization processing. Applying the reconstruction algorithm, the sparse signal x can be recovered from the observed signal y with a high probability. The typical reconstruction algorithm for recovering sparse signal x is divided into convex optimization algorithm and greedy algorithm. The convex optimization algorithm can obtain the sparse solution by minimizing the l_1 norm.

$$\hat{x} = \arg\min\|x\|_1; s.t. \quad y = \Phi x. \tag{2}$$

A typical convex optimization algorithm is the base tracking noise reduction algorithm (BP) [17]. The advantage of this algorithm is that the algorithm has low complexity and high reconstruction precision. However, in the face of a large amount of data, this algorithm takes too long to process and does not have real-time performance. The greedy algorithm converts the overall optimal solution into a local optimal solution each time to solve the problem. This type of reconstruction algorithm not only has a simple algorithm, but also has high reconstruction precision and can process a large amount of data in a short time. It mainly includes Matching Pursuit (MP) algorithm, OMP algorithm [18], Regularized OMP (ROMP) algorithm [19, 20], and Compressive sampling matching pursuit (CoSaMP) algorithm [21, 22], SP algorithm [23] and so on. Due to its fast convergence speed and low computational complexity, it has been widely studied in academia.

2.2 SP Algorithm

The difference between the SP algorithm and the OMP algorithm is mainly in the way of generating a set of reconstructed atomic vectors. The OMP algorithm and its improved algorithm select an optimal set of atomic vector sets for each iteration and then add it to the atomic index set. Add one atomic vector set per iteration until the iteration K times, reconstructing the channel from the iterated atomic vector set. In the SP algorithm, K optimal atomic vector sets are selected for each iteration, and the optimization is gradually refined by continuously updating the atomic vector set until the requirements are met. The idea of backtracking algorithm is used in the process of refinement, and the optimal atomic vector set can be systematically found. The above algorithm can reconstruct any sparse signal if the measurement matrix satisfies the RIP principle. It is widely used in practice. The implementation steps are as follows:

Input parameters: observation vector y, sensing matrix A, sparsity K.
Output parameter: K sparse approximation of h.

Initialization: The atomic index set T^0 is set to the index corresponding to the first K maximum values of $u = |\langle A, y \rangle|$. The residual y_r^0 is updated with the initialized atomic index set T^0.

$$y\_r^0 = y - A_{T^0}(A'_{T^0}A_{T^0})^{-1}A'_{T^0}y \tag{3}$$

Loop iteration, assuming the following is the m-th iteration loop.

Step 1: Update the atomic index set. The first K maximum index sets with the largest absolute value of the new residual and the inner matrix of the sensing matrix are:

$$T\_add = \max\{|\langle A, y\_r^{m-1}\rangle|, K\} \tag{4}$$

$$T^m = T^{m-1} \cup T\_add \tag{5}$$

Step 2: Estimating channel parameters using least squares;

$$\hat{h} = (A'_{T^m} * A_{T^m})^{-1} * A_{T^m} * y \tag{6}$$

Step 3: Update the atomic index set again;

$$T^m = \max(\left|\left\langle \hat{h}, E \right\rangle\right|, K) \tag{7}$$

Step 4: Update residual y_r^m with (6);
where E is an N-dimensional unit matrix.

Step 5: Judge: m is greater than K?

(a) Greater: exit loop
(b) Not greater: then m = m + 1 continues to execute Step 1.

From the traditional sparsely reconstructed SP algorithm, the sparsity K acts as a threshold decision condition and the atomic vector set number selection principle. It can be seen that the sparsity K is important in the whole algorithm. However, in ultra-wideband channel estimation, the sparsity of channel parameters is unknown and varies in real time. Considering the actual engineering application environment, this paper proposes an adaptive channel reconstruction algorithm with unknown sparsity.

3 UWB Channel Reconstruction Algorithm Based on Blind Sparsity

3.1 UWB Channel Estimation Model

The single-user UWB communication system of the PAM modulation system transmits signals is:

$$x(t) = \sum_{i=-\infty}^{+\infty} b_i p(t - iT_p) = \sum_{i=-\infty}^{+\infty} \sum_{j=0}^{N_s-1} b_{i_j} p(t - iT_p - jT_s) \tag{8}$$

where $p(t)$ is the transmit pulse, T_p is the period of each packet, T_s is the frame period, b_{i_j} is the i-th data symbol of the j-th data packet, and N_s is the number of data in one data packet. Referring to the IEEE802.15.4a channel model, the discrete UWB channel model in a complex multipath environment is [24]:

$$h(t) = \sum_{i=0}^{L-1} \alpha_i \delta(t - \tau_i) \tag{9}$$

L is the number of channel multipaths, δ is the unit impulse function, α_i is the fading of the i-th multipath, and τ_i is the delay of the i-th multipath. In the UWB system, $h(t)$ has sparsity. According to statistics, 10% of channel multipath accounts for 85% of the total channel energy [8].

The signal received by the receiving end after passing the UWB channel is:

$$r(t) = x(t) * h(t) + n(t) = \sum_{l=0}^{L-1} \alpha_l x(t - \tau_i) + n(t) \tag{10}$$

where * represents a convolution, $n(t)$ represents additive white Gaussian noise in the channel. The discrete form of Eq. (10) is expressed as:

$$r = xh + n \tag{11}$$

In order to avoid the measurement matrix Φ, the measurement compression of the noise n is performed. Add a FIR filter at the receiving end to filter out the noise n and observe it. The measured discrete form is:

$$y = \Phi r = \Phi x h \tag{12}$$

where x is the training sequence, so that $A = \Phi x$ (12) can be converted to:

$$y = Ah \tag{13}$$

The channel h of the UWB system is a sparse channel, and the mathematical models of the Eqs. (13) and (1) are identical. The UWB channel estimation problem can be converted to the problem of recovering the sparse channel h from the observation vector by using the compressed sensing reconstruction algorithm when A satisfies the RIP principle.

3.2 Adaptive Sparsity

For the input signal sparsity, this paper adaptively estimates the size of the old and new residuals. The signal is reconstructed without the need for signal sparsity as a priori knowledge. The main innovation of the algorithm is to set the initial pre-estimation sparsity to s, and select the s optimal atomic vector set in the observation matrix Φ. The size of s is n/10, and the sparsity of the general reconstructed signal is greater than s, y is projected onto $\Phi(vs)$ to get x_p, selecting the first s largest index value set in x_p, updating the residual y_r_n, and determining whether the new residual is smaller than the original residual. Less than the original residual, then s plus one continues to cycle. If it is larger than the original residual, it jumps out of the loop. The basic principle is to continuously increase the value of s. By comparing the mean of the old and new residuals to determine whether s is the sparsity K, the closer the value of s is to the sparsity K, the smaller the residual, only when s iterates to the sparsity K, The structure has the smallest residual. It should be noted that the pre-estimated sparsity must be smaller than the actual signal sparsity. To reduce the number of iterations, refer to [8] channel sparsity ratio, which is set to one tenth of the signal length. The specific adaptive process is shown in the Fig. 1.

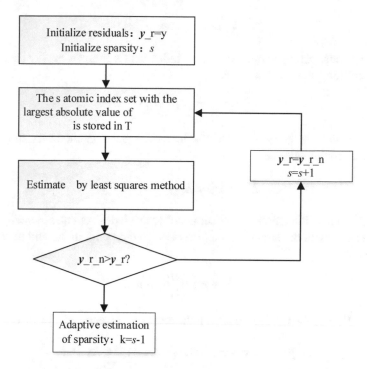

Fig. 1 Sparse degree adaptive schematic

3.3 The Proposed ASSP Algorithm Steps

For the traditional SP algorithm, the a priori signal sparsity K is needed to accurately reconstruct the signal with low signal residual. The specific steps of the adaptive sparsity subspace tracking algorithm are as follows.

The model is reconstructed by the formula (13). The core process of the detailed algorithm improvement is:

Input parameters: observation vector y, sensing matrix A.

Output parameters: estimate vector $\hat{h}$ and residual y_r.

Step 1: Data initialization:$y\_r = y$, atomic index set $T^0 = []$, current iteration number $t = 1$, initial atomic index set number s;

Step 2: The absolute value of the product of the residual y_r and the sensing matrix is stored in T_add according to the index value corresponding to the atom with the largest absolute value of the first s inner product in the order of largest to smallest, as in Eq. (15).

$$u = \{u_i | u_i = |\langle y\_r, \alpha_i \rangle|, i = 1, 2..., N\} \tag{14}$$

In Eq. (14), α_i is the ith column vector of the sensing matrix A.

$$T\_add = \max\{\boldsymbol{u}_i, s\} \tag{15}$$

Step 3: Constructing an atomic support index set $\boldsymbol{C}$: Merged by T_add and the last iteration atomic index value.

$$\boldsymbol{C} = T\_add \cup \boldsymbol{T}^{t-1} \tag{16}$$

Step 4: Calculate the estimated vector by least squares method according to the updated atomic index set.

$$\hat{\boldsymbol{h}} = (\boldsymbol{A}_{C^t} * \boldsymbol{A}_{C^t})^{-1} * \boldsymbol{A}_{C^t} * \boldsymbol{y} \tag{17}$$

Step 5: Update the atomic index set and find the index set corresponding to the first s atom with the largest absolute value of the product in the unit matrix and store it in.

$$\boldsymbol{T}^t = \max(\left\|\left\langle \hat{\boldsymbol{h}}, E \right\rangle\right\|, s) \tag{18}$$

Step 6: Update the residuals using all the atoms of the updated atom set in step 5.

$$\boldsymbol{y\_rn} = \boldsymbol{y} - \boldsymbol{A}_{T^t} * (\boldsymbol{A}_{T^t} * \boldsymbol{A}_{T^t})^{-1} * \boldsymbol{A}_{T^t} * \boldsymbol{y} \tag{19}$$

Step 7: Whether the iterative residual is satisfied, if it is satisfied, it stops the iterative output estimation vector $\hat{\boldsymbol{h}}$ and the residual. Otherwise $t = t + 1$ and $s = s + 1$, update residual $\boldsymbol{y\_r} = \boldsymbol{y\_rn}$ to proceed to step 2.

4 Performance Analysis

4.1 ASSP Algorithm Reconstruction Performance

In order to verify the reconstruction performance of the proposed ASSP algorithm, firstly, the reconstruction performances of the three algorithms SAMP, SP and ASSP are compared under the same sparsity. A set of data with a length of 256 and a sparsity of 20 is randomly generated by Gaussian function, and reconstructed by SAMP, SP and ASSP algorithms respectively. The reconstruction probability curves of the three algorithms under different measured values are shown in Fig. 2. It can be seen from the figure that under the condition of the same sparsity and signal length, the reconstruction probability of the ASSP algorithm is higher when the measured value is in the interval (60, 100). On the other hand, the critical measurement value of the ASSP algorithm is 90, which is lower than the SP algorithm (110). It shows that ASSP can accurately reconstruct the signal at a lower sampling rate, and the reconstruction performance of the ASSP algorithm is better than the other two algorithms.

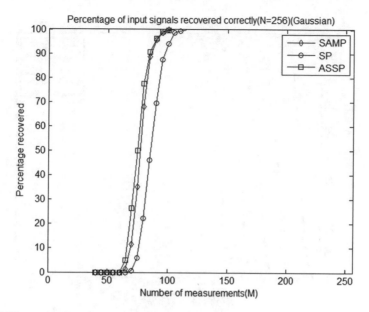

Fig. 2 Different algorithm reconstruction probability curve

Second, considering the reconstruction probability of the ASSP algorithm under different signal sparsity. Set the signal length to 256 and the signal sparsity to 4, 12, 20, 28 and 36 respectively. The relationship between the signal reconstruction rate and the measured value is shown in Fig. 3.

On the one hand, under the condition that the measured values are the same, the lower the sparsity, the greater the reconstruction probability. On the other hand, the greater the sparsity, the larger the measurement required to fully reconstruct the signal.

4.2 ASSP Algorithm Channel Estimation Performance Analysis

Set simulation parameters: The training sequence selects an identity matrix with a length of 300. The number of analog channel taps is 300, the number of measured values M is 70, and the sparsity is 20. The channel estimation result using the ASSP algorithm is compared with the original channel as shown in Fig. 4.

It can be seen from the figure that when the sampling rate is reduced to about one quarter of the traditional sampling rate, the residual of the estimated channel is 0.2331. The sampling rate is greatly reduced within the allowable residual tolerance range, which reduces the technical difficulty of hardware implementation.

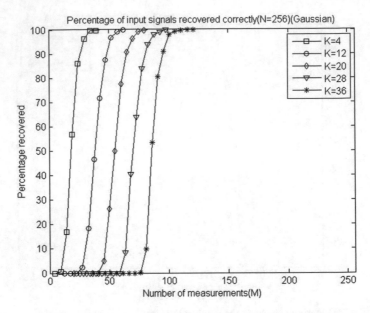

Fig. 3 Reconstruction probability curve of ASSP algorithm with different sparsity

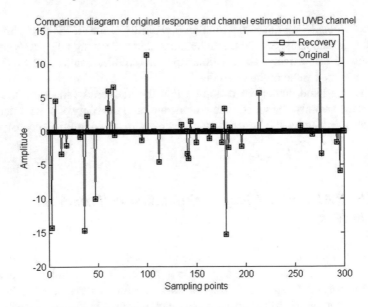

Fig. 4 ASSP algorithm reconstruction channel comparison chart

By comparing the channel estimation parameters with the original channel parameters, the proposed ASSP algorithm can basically realize the accurate reconstruction of the channel data. To further measure the performance of the channel estimation algorithm, it is evaluated by normalized mean square error. Its expression is:

$$MES = \frac{1}{N} \sum_{i=1}^{N} (\hat{h}_i - h_i)^2 \tag{20}$$

where $\hat{h}_i$ is the estimated parameter of the i-th path, and h_i is the original parameter of the i-th path.

For the data that the number of taps N is 256, the sparsity is 30, and the measured value M is 140, the simulation curves of the normalized mean square error of SP, SAMP, and ASSP with SNR change are shown in Fig. 5. The simulation uses the 802.15.4a channel model as the simulation environment. In order to ensure the uncorrelation between the observation matrix and the training sequence signal, the observation matrix adopts a Gaussian random matrix $A \in R^{M*N}$. In order to satisfy the RIP characteristics of the measurement matrix in the compressed sensing, the measurement matrix adopts a random Gaussian measurement matrix. The channel noise is additive white Gaussian noise. The simulation results show that the normalized mean square error performance of the ASSP algorithm proposed in this paper is better than the traditional compressed sensing SP algorithm and SAMP algorithm under the same measurement matrix and noise. SAMP, SP and ASSP algorithms increase with the signal-to-noise ratio. The training sequence provides more prior

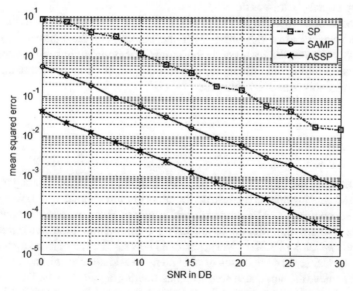

Fig. 5 Shows the MSE performance comparison of different algorithms

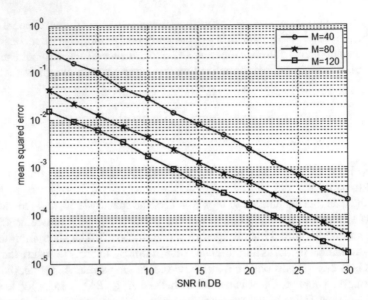

Fig. 6 Comparison of MSE performance of ASSP algorithm under different measured values

knowledge for channel estimation. The mean square error is gradually reduced until the signal-to-noise ratio reaches 30 dB.

The normalized mean square error simulation of the reconstructed channel using the ASSP algorithm for different observations is shown in Fig. 6. Under different observation conditions, the mean square error of the ASSP algorithm decreases with increasing signal-to-noise ratio.

It is not difficult to find by comparison. In the low SNR environment, the normalized mean square error decreases with the increase of the number of observations. This is due to the increase in the number of observations, which makes the pre-estimated information obtained by channel estimation more abundant, and the channel estimation accuracy is higher.

5   Conclusion

In order to solve the problem of UWB radio sampling difficulty, this paper uses the compressed sensing sparse reconstruction algorithm to estimate the UWB channel. A sparse reconstruction ASSP algorithm is proposed based on the SP reconstruction algorithm. The algorithm does not use sparsity as a priori condition, and initializes the sparsity to one tenth of n. Iteratively generates new residuals by gradually increasing the sparsity. Using the comparison of old and new residuals, the minimum residual is sought to achieve the purpose of adaptively estimating signal sparsity. The priori sparsity k is exchanged at the expense of reconstruction time and computation. Simulation experiments show that the reconstruction algorithm can reconstruct the channel with higher precision. It has important guiding significance for the application of practical engineering.

Acknowledgements This work has been supported by the following units and projects. They are the National Key R&D Program of China (2018YFB0505103), the National Natural Science Foundation of China (61561016, 61861008), Department of Science and Technology of Guangxi Zhuang Autonomous Region (AC16380014, AA17202048, AA17202033), Sichuan Science and Technology Plan Project (17ZDYF1495), Innovation Project of Guet Graduate Education(2018YJCX22) Guilin Science and Technology Bureau Project (20160202, 20170216), the basic ability promotion project of young and middle-aged teach-ers in Universities of Guangxi province (ky2016YB164), research on blind estima-tion of signal parameters for DSSS Communication (2019YCXS024).

References

1. Landolsi MA (2015) Signal design for improved multiple access capacity in DS-UWB communication. Kluwer Academic Publishers
2. Serikawa S, Lu H (2014) Underwater image dehazing using joint trilateral filter. Comput Electr Eng 40(1):41–50
3. Lu H, Li Y, Mu S, Wang D, Kim H, Serikawa S (2018) Motor anomaly detection for unmanned aerial vehicles using reinforcement learning. IEEE Internet of Things J 5(4):2315–2322
4. Lu H, Li Y, Chen M, Kim H, Serikawa S (2018) Brain Intelligence: go beyond artificial intelligence. Mobile Netw Appl 23:368–375
5. Lu H, Wang D, Li Y, Li J, Li X, Kim H, Serikawa S, Humar I (2019) CONet: a cognitive ocean network. IEEE Wireless Communications, in press
6. Lu H, Li Y, Uemura T, Kim H, Serikawa S (2018) Low illumination underwater light field images reconstruction using deep convolutional neural networks. Future Generat Comput Syst 82:142–148
7. Islam SMR, Kwak KS (2013) Preamble-based improved channel estimation for multiband UWB system in presence of interferences. Telecomm Syst 52(1):1–14
8. Cheng X, Wang M, Guan YL (2015) Ultra wideband channel estimation: A Bayesian compressive sensing strategy based on statistical sparsity. IEEE Trans Veh Technol 64(5):1819–1832
9. Ping W, Huailin R, Fuhua F (2014) UWB multi-path channel estimation based on CS-CoSaAMP algorithm. Comput Eng Appl 50(4):227–230
10. Huanan Y, Shuxu G (2012) Research on CS-based channel estimation methods for UWB communications. J Electron Inf Technol 34(6):1452–1456
11. Weidong W, Junan Y (2013) Ultra wide-band communication channel estimation based on Bayesian compressed sensing. J Circuits Syst 18(1):168–176
12. Mehmet R, Serhat K, Rpan HA (2015) Bayesian compressive sensing for ultra-wideband channel estimation: algorithm and performance analysis. Kluwer Academic Publishers
13. Cohen KM, Attias C, Farbman B et al (2014) Channel estimation in UWB channels using compressed sensing. In: IEEE international conference on acoustics, speech and signal processing. IEEE, pp 1966–1970
14. Fuhua F, Huailin R (2014) Non-convex compressive sensing utral-wide band channel estimation method in low SNR conditions. Acta Electronica Sinica 42(2):353–359
15. Yanfen W, Xiaoyu C, Yanjing S (2017) Sparsity adaptive algorithm for ultra-wideband channel estimation. J Univ Electron Sci Technol China 46(3):498–504
16. Candès EJ, Romberg JK, Tao T (2010) Stable signal recovery from incomplete and inaccurate measurements. Commun Pure Appl Math 59(8):1207–1223

17. Xianyu Z, Yulin L, Kai W (2010) Ultra wide-band channel estimation and signal detection through compressed sending. J Xi'an Jiao Tong Uni 44(2):88–91
18. Tropp JA, Gilbert AC (2007) Signal recovery from random measurements via orthogonal matching pursuit. IEEE Trans Inf Theory 53(12):4655–4666
19. Schwab H (2007) Signal recovery from incomplete and inaccurate measurements via regularized orthogonal matching pursuit. Submitted for publication. IEEE J Select Topics Signal Process 4(2):310–316
20. Needell D, Vershynin R (2010) Signal recovery from incomplete and inaccurate measurements via regularized orthogonal matching pursuit. IEEE J Select Topics Signal Process 4(2):310–316
21. Davenport MA, Needell D, Wakin MB (2013) Signal space CoSaMP for sparse recovery with redundant dictionaries. IEEE Trans Inf Theory 59(10):6820–6829
22. Zhang L (2015) Image adaptive reconstruction based on compressive sensing via CoSaMP. Appl Mech Mater 631–632:436–440
23. Dai W, Milenkovic O (2009) Subspace pursuit for compressive sensing signal reconstruction. IEEE Trans Inf Theory 55(5):2230–2249
24. Lei S, Zheng Z, Liang T (2012) Ultra wideband channel estimation based on kalman filter compressed sensing. Trans Beijing Instit Technol 32(2):64–67+77

A New Unambiguous Acquisition Algorithm for BOC(n, n) Signals

Xiyan Sun, Qing Zhou, Yuanfa Ji, Suqing Yan, and Wentao Fu

Abstract The autocorrelation function of the BOC modulated signal has more than one peak, which causes the receiver to capture the error signal. Therefore, this paper proposes a new fuzzy-free capture algorithm for BOC(n, n) signals. Divide the local BOC signal into two branch signals. Then, correlating the tributary signal with the received BOC signal, and the correlation function is then shifted by a quarter and three quarters of the chip, respectively, for reconstructed. The simulation analysis shows that the algorithm can weaken interference of the secondary peak and has a small amount of calculation. Both capture sensitivity and de-blurring performance are improved.

Keywords Boc(n, n) · Multiple peaks · Correlation · Unambiguous acquisition

1 Introduction

In order to ensure that the various modern GNSS (Global Navigation Satellite System) systems [1, 2] can work in different frequency bands, the spectral splitting characteristics of BOC (Binary Offset Carrier) [3] become an opportunity to use it as a navigation satellite signal. In addition, because BOC has higher positioning accuracy than traditional BPSK (Binary phase shift keying), various BOC modulations become an important component of modern GNSS systems and become the main candidate for the development of GNSS systems. Nevertheless, the multi-peak nature of autocorrelation function of BOC signal causes the receiver to have severe

X. Sun · Q. Zhou · Y. Ji (✉) · S. Yan · W. Fu
Guangxi Key Laboratory of Precision Navigation Technology and Application, Guilin University of Electronic Technology, Guilin 541004, China
e-mail: jiyuanfa@163.com

National & Local Joint Engineering Research Center of Satellite Navigation and Location Service, Guilin 51004, China

X. Sun · Y. Ji · S. Yan · W. Fu
Guangxi Experiment Center of Information Science, Guilin University of Electronic Technology, Guilin 541004, China

© Springer Nature Switzerland AG 2021
H. Lu (ed.), *Artificial Intelligence and Robotics*,
Studies in Computational Intelligence 917,
https://doi.org/10.1007/978-3-030-56178-9_19

ambiguities in its baseband signal processing, which can cause positioning bias that cannot be ignored. Therefore, it is a key issue for research to eliminate the ambiguity of correlation peaks.

The solutions currently proposed mainly include the following: The BPSK-LIKE algorithm [4] is further divided into a single sideband method [5, 6] and a double sideband method. In [7], the BPSK-LIKE algorithm captures local pseudocodes after moving them to both sides of the spectrum. Although BPSK-like can remove the ambiguity of BOC and has a simple structure, the disadvantage is that the correlation peak similar to BPSK is wider than the BOC signal in terms of correlation peaks; The SPCP [8, 9] algorithm simulates the method of quadrature demodulation stripping carrier, and uses a pair of mutually orthogonal local BOC codes to remove the influence of subcarriers on BOC signal acquisition, which is similar to the capture result obtained by BPSK-like technology. The essence of the ASPeCT [10, 11] algorithm is that the signal received by the receiver is related to the local BOC code, and the signal is correlated with the PRN code, and then reconstructed by using the above two correlation functions to eliminate the influence of the subcarrier. This method successfully suppresses the influence of subcarriers and can improve the acquisition accuracy, but the method is computationally intensive and only works for the BOC(n, n) group. The BOC signal direct processing method [12] improves the peak-to-peak ratio of the primary and secondary peaks. Although the algorithm is simple to implement, it cannot handle the ambiguity. Reference [13] also obtains the side peak elimination capability by constructing a local auxiliary signal, but only for high-order BOC signals. A variety of capture and tracking techniques [14–18] have been proposed for the ambiguity of the BOC capture process.

In this article, a new improved algorithm is proposed. Firstly, the local BOC signal is split, and one set of odd-branch signals is correlated with the received signal, and then shifted and combined to achieve the purpose of eliminating the peaks (Alternatively, the even tributary signal is correlated with the received signal, shifted and recombined to eliminate side peaks). This paper takes the odd branch signal as an example to explain. The simulation shows that the algorithm effectively eliminates the secondary peak of the correlation function and reduces the main peak span to three quarters of the chip, which can result in sharp and narrow correlation peaks for better acquisition performance.

2 Fuzziness Problem

This article studies the BOC(n, n) group signals used by most modern GNSS navigation systems. Define the product of the PRN and the subcarrier as the BOC code, denoted by $S_{boc}(t)$ as:

$$S_{boc}(t) = c(t)sc(t) \tag{1}$$

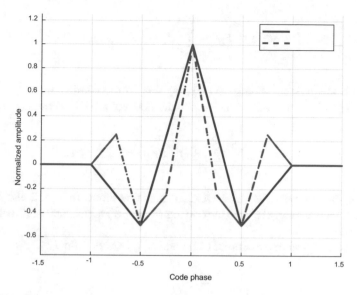

Fig. 1 BOC(1, 1) autocorrelation function

where $c(t)$ is the PRN (Pseudo Random Noise) code; $sc(t)$ represents the binary subcarrier.

This article describes the BOC(1, 1) signal, the same characteristics of BOC(n, n). Figure 1 shows the autocorrelation function of BOC(1, 1). It can be seen from the figure that there is a secondary peak in the BOC signal, which will cause blurring during acquisition problem.

3 Proposed Method

3.1 Algorithm Analysis

This paper proposes a new no-fuzzy acquisition algorithm to solve the ambiguity problem. First, split the BOC signal. The pseudo-random code can be modeled as:

$$c(t) = \sum_{i=-\infty}^{\infty} C_i P_{Tc}(t - i T_C) \tag{2}$$

where C_i is a chip value, $C_i \in (-1, 1)$; P_{Tc} is one rectangular pulse; T_c is the width of a chip. The local sub-carrier is expressed as:

$$sc(t) = P_{Tc} = \sum_{j=0}^{N-1} d_j \, P_{T_{sc}}(t - j \, T_{sc}) \tag{3}$$

where d_j is the pulse symbol, $d_j \in (-1, 1)$; N is the modulation order.

Take (1) and (2) into (3), the mathematical model of the BOC baseband signal is:

$$S_{boc}(t) = \sum_{j=0}^{N-1} \sum_{i=-\infty}^{\infty} C_i \, d_j \, P_{T_{sc}}(t - i \, T_C - j \, T_{sc}) \tag{4}$$

The separation process of the BOC(1, 1) signal is shown in Figs. 2 and 3, which $C_e(t)$ represents the signal part of the odd road and $C_o(t)$ represents the signal portion of the even road.

The mathematical expression of $C_e(t)$ and $C_o(t)$ are as follows:

$$C_e(t) = \sum_{i=0}^{N_c-1} d_e \, C_i \, P_{T_{SC}}(t - i \, T_C) \tag{5}$$

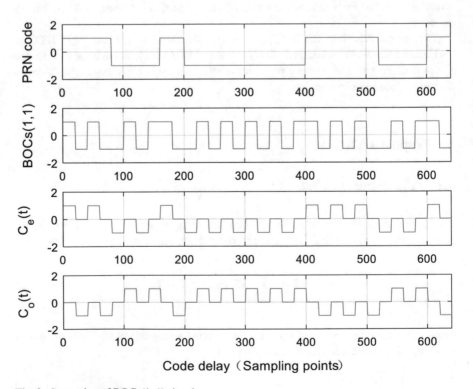

Fig. 2 Separation of BOCs(1, 1) signals

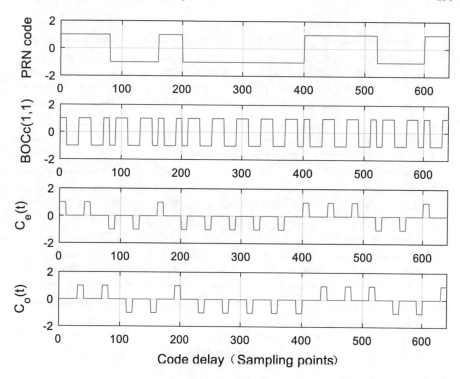

Fig. 3 Separation of BOCc(1, 1) signals

$$C_o(t) = \sum_{i=0}^{N_c-1} d_o\, C_i\, P_{Tsc}\,(t - i\, T_C - T_{sc}) \tag{6}$$

where N_c is the number of PRN chips in a certain period of time; d_e is a subcarrier pulse symbol of the odd branch signal; d_o is a subcarrier pulse symbol of the even branch signal, and $d_o\ \&\ d_E \in d_j$.

The reconstruction rule of the algorithm in this paper is:

$$R = |R_{e1} - R_{e2}| - |R_{e1} + R_{e2}| \tag{7}$$

R_{e1} is a new correlation function that shifts $1/4$ chip to the left, and R_{e2} is a new correlation function that shifts $3/4$ chip to the right. The reconstructed correlation function is shown in Figs. 4 and 5.

Figures 4 and 5 show that the unambiguous correlation function can be obtained through the reconstruction rule in (7).

The IF signal expression of BOC(n, n) is:

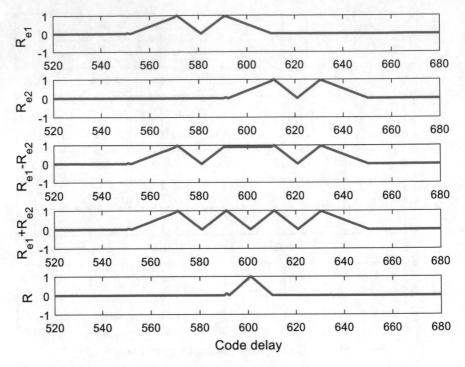

Fig. 4 BOCs(1, 1) correlation function reconstruction

$$r(t) = \sqrt{P_S} \times C(t - \tau) \times D(t - \tau) \times SC(t - \tau) \times \cos(2\pi(f_{IF} + f_D)t) + n(t) \tag{8}$$

P_S is the power of the input signal, $D(t)$ is the navigation data, τ is the code delay, and $n(t)$ is the noise term [19].

The input signal is mixed with the local carrier and multiplied by the odd-branch signal to obtain:

$$r_e(t) = r(t)[\cos[2\pi(f_{IF} + f_D)t] + j\sin[2\pi(f_{IF} + f_D)t]]C_e(t) + n_e \tag{9}$$

Then ahead of the $1/4$ chip to get:

$$r_{e1}(t) = r(t)[\cos[2\pi(f_{IF} + f_D)t] + j\sin[2\pi(f_{IF} + f_D)t]]C_e\left(t + \frac{T_C}{4}\right) + n_{e1} \tag{10}$$

Lagging it the $3/4$ chip can be obtained:

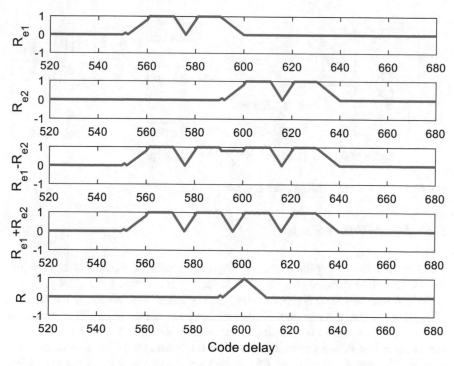

Fig. 5 BOCc(1, 1) correlation function reconstruction

$$r_{e2}(t) = r(t)[\cos[2\pi(f_{IF} + f_D)t] + j\sin[2\pi(f_{IF} + f_D)t]]C_e\left(t - \frac{3T_C}{4}\right) + n_{e2}$$

$$(11)$$

Perform coherent integration for (10) to get:

$$\bar{r}_{e1}(t) = P_S R_{e1}(\Delta\tau)T_s \sin c(\pi\Delta f_D T_s) \times [\cos(\pi\Delta f_D T_s) + \sin(\pi\Delta f_D T_s)] + N_{e1}$$

$$(12)$$

Perform coherent integration for (11) to get:

$$\bar{r}_{e2}(t) = P_S R_{e2}(\Delta\tau)T_s \sin c(\pi\Delta f_D T_s) \times [\cos(\pi\Delta f_D T_s) + \sin(\pi\Delta f_D T_s)] + N_{e2}$$

$$(13)$$

$R_{e1}(\Delta\tau)$ and $R_{e2}(\Delta\tau)$ are new correlation functions of shifting $1/4$ chip to the left and shifting $3/4$ chip to the right, respectively, Δf_D is the Doppler frequency offset, N_{e1} and N_{e2} are Gaussian white noises.

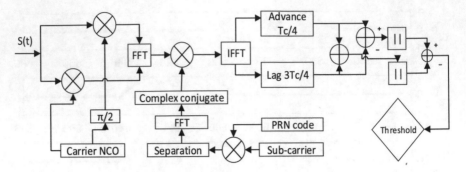

Fig. 6 New algorithm acquisition block diagram

3.2 Algorithms Scheme

The algorithm block diagram is shown in Fig. 6. According to the flow in the drawing, the carrier is first stripped out, and then subcarrier modulated local PRN sequence is divided into two branch signals with the subcarrier pulse length as a reference. This paper focuses on the related shifting mode of odd-branch signal. The correlation between the two signals is obtained by correlating the branch signal and the input signal described in this paper. Then move the left side by $1/4$ chip and the right side by $3/4$ chip to get two new functions. Then the two new functions are added, subtracted, and modulo subtracted, and the edge of the autocorrelation function is finally eliminated.

4 Acquisition Performance Analysis

4.1 Acquisition Result Simulation

The sampling frequency is set to 40.92 MHz, so one spreading chip corresponds to 40 sampling points, and the setting code phase is the 601th sampling point, that is, the 16th chip. Figure 7 is a diagram showing the result of capturing the BOC(1, 1) signal.

In the matlab platform, the same parameters as shown in Fig. 7 are set. According to the Fig. 8, for the BOCs(1, 1) signal, the captured two-dimensional maps of the ASPeCT and SCPC methods have secondary peaks, and the algorithm completely eliminates the secondary peaks. And the correlation peak is very narrow. In Fig. 9, the algorithm is also superior to the SCPC algorithm for the BOCc(1, 1) signal.

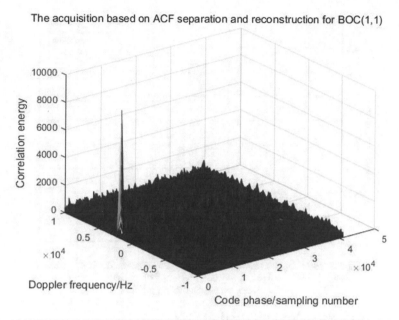

Fig. 7 Capture result graph of the new algorithm

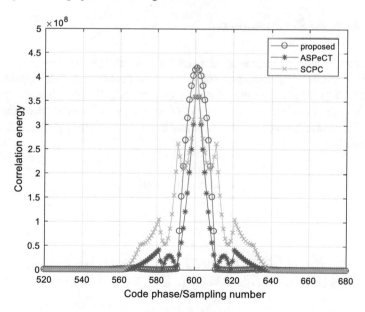

Fig. 8 Two-dimensional capture of BOCs(1, 1) signal

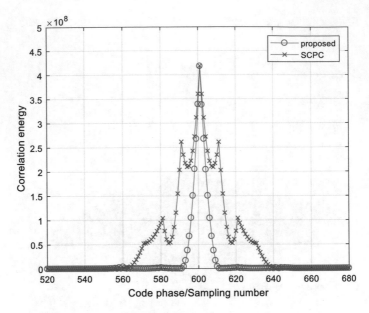

Fig. 9 Two-dimensional capture of BOCc(1, 1) signal

4.2 Detection Performance

The algorithm proposed judges whether the BOC signal is accurately catched based on the comparison of the detection statistic and the detection threshold value set by the decider. If the detected amount is greater than the detection threshold, the signal is accurately captured; if the detected amount does not exceed the detection threshold, the signal is not accurately captured. The traditional non-coherent detection statistics are as follows:

$$D = \sum_{j=1}^{M} (I_j^2 + Q_j^2) \qquad (14)$$

In (14),

$$\begin{cases} I_j = \sqrt{T_s C/N_0} \sin c(\pi \Delta f_D T_s) R(\Delta \tau) \cos(\Delta \varphi) + N_{i,j} \\ Q_j = \sqrt{T_s C/N_0} \sin c(\pi \Delta f_D T_s) R(\Delta \tau) \sin(\Delta \varphi) + N_{q,j} \end{cases} \qquad (15)$$

where M is accumulative times, C/N_0 is the carrier-to-noise ratio, $\Delta \varphi$ is carrier phase error, $R(\Delta \tau)$ represents a correlation function obtained by a correlation operation between the local signal and the received signal, $N_{i,j}$ and $N_{q,j}$ are Gaussian white noise.

Bring the formula (15) into the formula (14), D can be expressed as:

$$D = \sum_{j=1}^{M} \left[\begin{array}{l} (\sqrt{T_s C/N_0} \sin c(\pi \Delta f_D T_s) R(\Delta \tau) \cos(\Delta \varphi) + N_{i,j})^2 \\ +(\sqrt{T_s C/N_0} \sin c(\pi \Delta f_D T_s) R(\Delta \tau) \sin(\Delta \varphi) + N_{q,j})^2 \end{array} \right] \quad (16)$$

Non-central parameter ξ^2 of D is:

$$\xi^2 = M T_s C/N_0 \sin c^2(\pi \Delta f_D T_s) R^2(\Delta \tau) \quad (17)$$

It can be seen from Eqs. (16) and (17) that when the $\Delta \tau$ is smaller until it is reduced to zero, the statistical detection amount D can be maximized.

Assuming $P_D(x)$ is the probability density, then the detection probability P_d at this time is:

$$P_d = \int_{V}^{+\infty} P_D(x)dx \quad (18)$$

where V is the capture threshold.

Simplify (12) and (13) to:

$$\bar{r}_{e1}(t) = r_{e1}(\Delta \tau, \Delta f_D) + N_{e1} \quad (19)$$

$$\bar{r}_{e2}(t) = r_{e2}(\Delta \tau, \Delta f_D) + N_{e2} \quad (20)$$

According to the reconstruction rule, r_{e1-e2} and r_{e1+e2} can be expressed as:

$$r_{e1-e2} = [r_{e1}(\Delta \tau, \Delta f_D) + N_{e1}] - [r_{e2}(\Delta \tau, \Delta f_D) + N_{e2}] \quad (21)$$

$$r_{e1+e2} = [r_{e1}(\Delta \tau, \Delta f_D) + N_{e1}] + [r_{e2}(\Delta \tau, \Delta f_D) + N_{e2}] \quad (22)$$

Take Eqs. (21) and (22) into formula (14), the new detection statistic D_1 can be expressed by the following formula:

$$D_1 = \sum_{j=1}^{M} ((r_{e1-e2})^2 + (r_{e1+e2})^2) \quad (23)$$

Now assuming that the probability density is $P_{D_1}(x)$, the detection probability P_{d1} is indicated as:

$$P_{d1} = \int_{V}^{+\infty} P_{D_1}(x)dx \quad (24)$$

The detection probability is simulated by MonteCarlo method which can perform performance analysis. A comparison of the detection probabilities of the BOC(1, 1) signals under several catch methods is shown in Figs. 10 and 11.

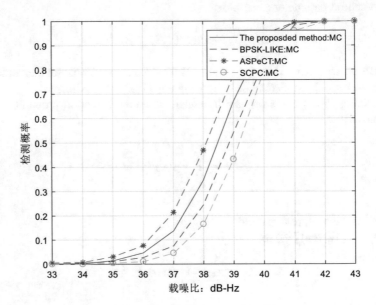

Fig. 10 Detection probability of BOCs(1, 1) signal

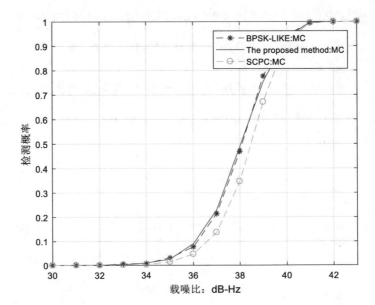

Fig. 11 Detection probability of BOCc(1, 1) signal

Table 1 The operations comparison of the three algorithms

Number	Algorithm name	Related times	Square times	Shift times
1	ASPeCT	2	2	0
2	SCPC	2	2	0
3	The proposed method	1	2	2

In Fig. 10, the detection performance of this algorithm is higher than SCPC and BPSK-LIKE for the sin-BOC(1, 1) signal. In Fig. 11, for the cos-BOC(1, 1) signal, the detection performance of the proposed algorithm is not much different from the BPSK-LIKE algorithm, but higher than the SCPC algorithm.

4.3 Algorithmic Computational Analysis

It can be seen from Table 1 that the ASPeCT algorithm needs to perform two correlations and squares separately; the calculation amount of the SCPC method is the same as that of the ASPeCT method; however, the proposed algorithm only needs to be correlated once, respectively shifted and squared twice, so which complexity is lesser.

5 Conclusion

This paper proposes a new improved method based on the related shift, which mainly uses BOC(1, 1) signal for theoretical analysis and simulation, and actually conforms other BOC(n, n) groups. Compared with the BPSK-LIKE and SPCP methods, proposed method has higher deblurring validity and higher capture sensitivity. erefore, the unambiguous acquisition method proposed in this paper is a good ice for modern GNSS receivers. The fuzzy acquisition method of other BOC ups has reference significance.

Acknowledgements This work has been supported by the following units and projects. They are e National Key R&D Program of China (2018YFB0505103), the National Natural Science Foun-tion of China (61561016, 61861008), Department of Science and Technology of Guangxi Zhuang utonomous Region (AC16380014, AA17202048, AA17202033), Sichuan Science and Tech-ology Plan Project (17ZDYF1495), Guilin Science and Technology Bureau Project (20160202, 20170216), the basic ability promotion project of young and middle-aged teachers in Universities of Guangxi province (ky2016YB164), the graduate education innovation program funding project of Guilin University of Electronic Technology (2019YCXS024).

References

1. Deng Z, Yue X, Lu Y (2015) Unambiguous sine-phased BOC(kn, n) signal acquisition based on combined correlation functions. Telkomnika 13(2):502
2. Li, Pun C, Xu F, Pan L, Zong R, Gao H, Lu H (2020) A hybrid feature selection algorithm based on an discrete artificial bee colony for Parkinson's diagnosis. ACM Trans Internet Technol
3. Hofmann-Wellenhof B (ed) (2007) GNSS—global navigation satellite systems: GPS, GLONASS, Galileo, and more. Springer-Wien, New York
4. Burian A, Lohan ES, Renfors M (2006) BPSK-like methods for hybrid-search acquisition of Galileo signals. In: IEEE International conference on communications. IEEE, pp 5211–5216
5. Zhou J, Liu C (2015) Joint data-pilot acquisition of GPS L1 civil signal. In: International conference on signal processing. IEEE, pp 1628–1631
6. Burian A, Lohan ES, Renfors M (2006) BPSK-like methods for hybrid-search acquisition of Galileo signals. In: Proceedings of the IEEE ICC, June, 2006, pp 5211–5216
7. Wang XJ, Lu MY, Li RH (2017) A single sideband BPSK-like acquisition algorithm for BOC(10, 5) signals. Inf Comput (Theor Ed) 12:86–88
8. Ward PW (2003) A design technique to remove the correlation ambiguity in binary offset carrier (BOC) spread spectrum signals. In: Proceedings of annual meeting of the Institute of Navigation & Cigtf Guidance test symposium
9. Yang Z, Huang Z, Geng S (2009) Unambiguous acquisition performance analysis of BOC(m, n) signal. In: International conference on information engineering and computer science. IEEE, pp 1–4
10. Julien O, Macabiau C, Cannon ME et al (2007) ASPeCT: unambiguous sine-BOC(n, n) acquisition/tracking technique for navigation applications. IEEE Trans Aerosp Electron Syst 43(1):150–162
11. Qian S, Yin X (2016) Research on ASPeCT-based acquisition and tracking of BOC modulating signal. Mod Electron Tech
12. Li P, Gao F, Li Q (2015) An improved unambiguous acquisition scheme for BOC(n, n) signals. In: International conference on wireless communications and signal processing. IEEE, pp 1–6
13. Yan T, Wei J, Tang Z et al (2015) Unambiguous acquisition/tracking technique for high-order sine-phased binary offset carrier modulated signal. Wirel Pers Commun 84(4):1–23
14. Wendel J, Schubert FM, Hager S (2015) A robust technique for unambiguous BOC tracking. Navigation 61(3):179–190
15. Feng T, Kai Z, Liang C (2016) Unambiguous tracking of BOC signals using coherent combination of dual sidebands. IEEE Commun Lett 20(8):1555–1558
16. Yuan-Fa JI, Liu Y, Zhen WM et al (2017) An unambiguous acquisition algorithm based on correlation for BOC(n,n) signal. IEICE Trans Commun E100.B(8)
17. Cao XL, Guo CJ (2016) A new unambiguity acquisition algorithm for BOC(n, n) signal. Position Syst 41(6):1–5
18. Sun X, Zhou Q, Ji Y et al. An unambiguous acquisition algorithm for BOC (n, n) signal based on sub-correlation combination. Wirel Pers Commun 1–20
19. Ren J, Yang G, Jia W et al (2014) Unitary unambiguous tracking method based on combined correlation functions for BOC modulated signals. Acta Aeronautica Et Astronautica Sinica 35(7):2031–2040

Printed in the United States
by Baker & Taylor Publisher Services